1933

◆ Official ◆

Auto-Radio
Service Manual

Complete Directory
of All
Automobile Radio Receivers

Full Installation and
Trouble Shooting Guide

Volume No. 1

Hugo Gernsback
Editor

Robert Hertzberg
Associate Editor

GERNSBACK PUBLICATIONS, INC.
PUBLISHERS
96-98 PARK PLACE, NEW YORK, N. Y.

ISBN: 1450596940

EAN: 9781450596947

This Is a New Index for the Entire Manual

ACME RADIO MFG. CO.
MOTO MIDGET

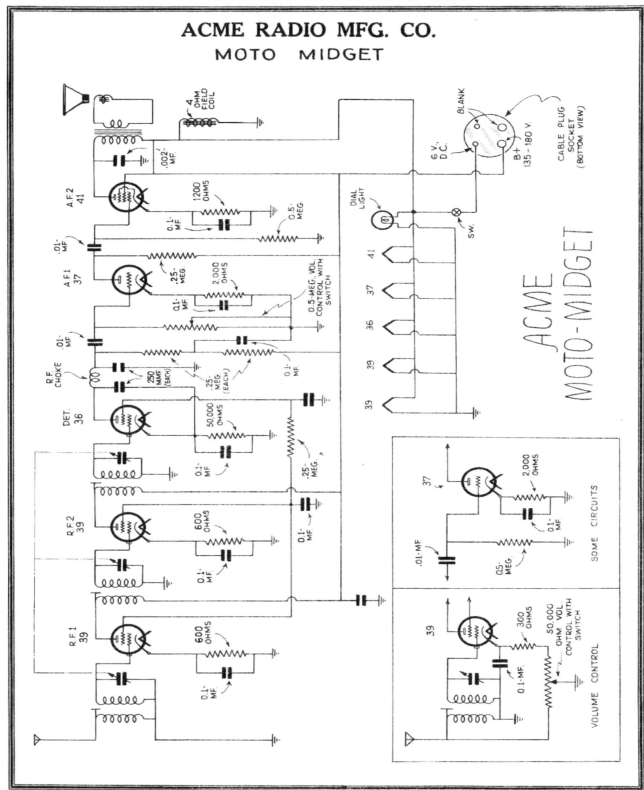

INDEX

ATWATER KENT MODELS 636 AND 756

MODELS 636 and 756 are similar six-tube superheterodynes equipped with all-electric power supply, eliminating the use of dry batteries. Model 636 is designed for attachment to and operation from the instrument panel of the car, the dynamotor power unit being in a separate metal container mounted under the floor, the front seat, or in any other convenient place. Model 756 is intended for cars on which the 636 cannot be mounted by reason of car design. In this type a special steering post control is provided and the radio chassis (complete with power supply in a single case) may be mounted under the floor board. With both models the loudspeaker is a separate unit.

Ignition suppressors included with set.

The tubes used are: two 41's, two 39's, one 36, and one 85. Among circuit features are automatic volume control, push-pull output, two-point

ATWATER KENT MFG. CO.

tone control. Tuning range 1500-540 kc. Tuning dial calibrated in kilocycles. Total battery consumption, including dynamic speaker, 5¾ amperes. I. F., 262½ kc.

AUTOCRAT MODEL TRF-41

A feature of this set is the small dimensions of the chassis which measures only 5¼ x 6½ x 8¼ ins. Police calls may be received on this set. Complete equipment, illustrated, at the list price includes: one shielded T.R.F. chassis with tubes, remote control unit, 30-foot cables, separate dynamic reproducer, sub-car antenna, six spark plug suppressors, distributor suppressor, distributor filter condenser, and miscellaneous hardware. The "B" supply, which may be batteries or an

eliminator, is an accessory to be bought separately.

The reproducer is of one-hole mounting type; the set, 3-hole. These two units are electrically connected by means of plugs. The receiver is designed for bulkhead mounting—as far as possible from the ignition coil. The remote-control unit clamps to the steering column. The sub-car antenna is to be slung between the front and rear axles, toward the rear. Place the "B" supply wherever convenient.

This T.R.F. receiver incorporates: two type 39 R.F. tubes; one 36, detector: one 37, first A.F.; and 41's in push-pull A¹, second A.F.; sensitivity, less than 1 microvolt-per-meter; power output, 3 watts. The total "B" voltage required is 180.

AUTOCRAT RADIO CO.

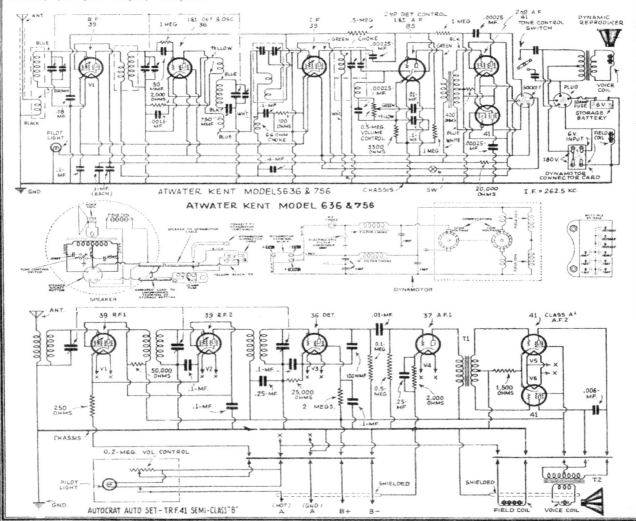

ATWATER KENT MODELS 636 & 756

ATWATER KENT MODEL 636 & 756

AUTOCRAT AUTO SET - T.R.F. 41 SEMI-CLASS "B"

AUTOMATIC RADIO MFG. CO.

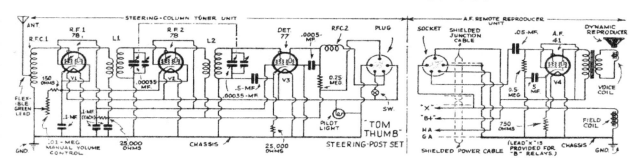

"TOM THUMB" STEERING POST SET

To MEET the demand for a quickly installed automotive radio set the Automatic Radio Mfg. Co. has added to their "Tom Thumb" line a Steering Post Auto Radio Set; this instrument and the accessories which are supplied to complete the ensemble are illustrated. Although the operating system includes four tubes, only three—the two R.F. amplifiers and the detector—are incorporated in the unit which mounts on the steering column (holes in the tuner case permit either left- or right-side mounting). The case of this "tuner" measures about 5x5½x5½ ins. deep. The remaining tube, the A.F. amplifier, is contained in the dynamic reproducer chassis, as shown in one of the views; the case of this component measures 4½x8 ins. diameter (about 10 ins., over all). These two units connect together by means of a cable which plugs into the tuner.

A complete installation at the list price includes: a steering-column tuner (with illuminated, full-vision dial, volume control, and off-on switch), an amplifier-reproducer, shielded junction cable and plug, shielded power cable, steering-column clamp, tubes, spark plug suppressors, distributor suppressor, generator bypass condenser, and hardware. Not included in the factory kit are the antenna and the "B" unit; either batteries, or any of the conventional "B" units with a voltage output within the range of 135 to 180 volts, may be used as power supply.

Clamping the tuner to the steering

post eliminates the necessity for running-control wires to a remote-control unit. By removing from the tuner chassis the L-shaped cover-plate, or from the reproducer chassis the cover-disc, the interior of these two units is available for replacing tubes—without the necessity of removing either of these chasses. The amplifier-reproducer chassis brackets may be bent to conform with irregular bulkheads or floorboards.

Referring to the diagram it is seen that this diminutive set, a literal "Tom Thumb" in automotive radio receivers, incorporates the following tube arrangement: one type 78 tube as an untuned R.F. amplifier; one 78 tuned R.F. amplifier; one 77, detector; one 41, A.F. amplifier. The total "B" current consumption is 28 milliamperes at 180 volts, or 18 at 135; the "A" drain (tube heaters and reproducer field) is 2.7 amperes; the total storage battery drain for "A" and "B" (using a good, standard "B" unit) is less than 4 amperes.

The Tom Thumb set has a sensitivity of about 4.5 microvolts at 1400 kc., about 40 at 750 kc., and about 20 at 600 kc.

The power cable has three main leads: hot, or ungrounded "A" (HA, in diagram), a "B" positive lead which taps to a 135 to 180 volt current source (the higher voltage will result in improved volume and tone quality), and a shield lead GA for the "B" negative and grounded "A." The leads from the dynamic reproducer are shielded. A special terminal, X, is provided in the reproducer and is connected to the

"live" side of the heaters. This wire is required for controlling the "B" supply or motor-generator unit which is not equipped with relays. A shielded lead run from this terminal to the supply device, and grounded at each end, will enable the "B" unit to be turned on or off with the switch on the front of the set.

Many Service Men prefer to use a dynamotor rather than batteries or an "eliminator" to supply the required "B" voltage.

The dynamotor shaft *must* be placed in a horizontal position.

Check the polarity of the dynamotor leads by means of a voltmeter. If the precaution of pre-determining the polarity of the dynamotor leads is not followed, the shielding of the hot "B" lead, being grounded, will short the output circuit of the dynamotor and perhaps permanently damage it; or, the set will not operate. If the leads to the storage battery are reversed, the storage battery will be shorted, due to the hot "A" lead shield being grounded, and the car battery thus may be permanently damaged.

To determine whether interference is due to noise pick-up by the flat-top or lead-in sections of the antenna system, remove the lead-in from the antenna, or substitute for the lead-in a short length of wire running in the same direction as, and the same length as, the regular lead-in. If the noise stops, it indicates that the flat-top portion of the antenna is picking up the interference.

PHILCO RADIO & TELEVISION CORP.

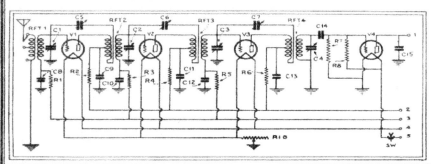

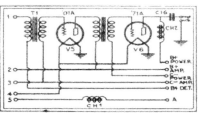

"Model NR. 107" tuning unit, left; "Model NR. 108" amplifier above. Either R7 or R8, or both, may be used. Observe resistance-capacity filters in grid and plate leads. Part values are not given.

CARTER GENEMOTOR CORP.

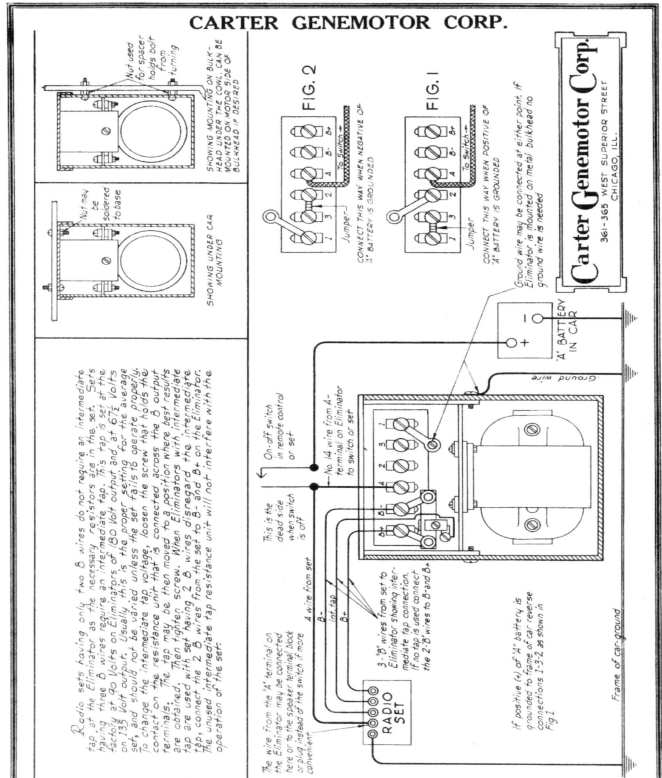

Carter Genemotor Corp.

361-365 WEST SUPERIOR STREET
CHICAGO, ILL.

FIG. 2

CONNECT THIS WAY WHEN NEGATIVE OF "A" BATTERY IS GROUNDED

Jumper

To Switch

FIG. 1

CONNECT THIS WAY WHEN POSITIVE OF "A" BATTERY IS GROUNDED

Jumper

To Switch

Ground wire may be connected at either point. If Eliminator is mounted on metal bulkhead no ground wire is needed

'A' BATTERY IN CAR

Ground wire

On-off switch in remote control or set

No. 14 wire from A- terminal on Eliminator to switch or set

This is the dead side when switch is off

A wire from set

B-

Int tap

B+

3- B wires from set to Eliminator showing inter- mediate tap connection. If no tap is used connect the 2 B wires to B and B+

RADIO SET

The wire from the "A" terminal on the Eliminator may be connected here or to the speaker terminal block or plug instead of the switch if more convenient

If positive (+) of "A" battery is grounded to frame of car reverse connections 1-3-2 as shown in Fig 1

Frame of car-ground

Nut used for spacer holds bolt from turning

SHOWING MOUNTING ON BULK- HEAD UNDER THE COWL. CAN BE MOUNTED ON MOTOR SIDE OF BULKHEAD IF DESIRED

Nut may be soldered to base

SHOWING UNDER CAR MOUNTING

Radio sets having only two B wires do not require an intermediate tap at the Eliminator as the necessary resistors are in the set. Sets having three B wires require an intermediate tap. This tap is set at the factory at 90 Volts on Eliminators of 180 Volt output and at 67½ Volts on 135 Volt output. Usually this is the proper setting for the average set, and should not be varied unless the set fails to operate properly. To change the intermediate tap voltage, loosen the screw that holds the contact on the resistance unit that is connected across the B output terminals. The tap may be then moved to a position where best results are obtained. Then tighten screw. When Eliminators with intermediate tap are used with set having 2 B wires disregard the intermediate tap, connect the 2 B wires from the set to B- and B+ on the Eliminator. The unused intermediate tap resistance unit will not interfere with the operation of the set.

CROSLEY RADIO CORP.

Model 95 (Roamio)

Specifications

Model 95 is a compact automobile receiver measuring only 10¾-in. long by 4½-in. wide by 6⅛-in. high. It may be mounted on the floor of the automobile, on the engine bulkhead, or in any other convenient location. Normally it obtains its power from the automobile storage battery and dry "B" batteries. A "B" eliminator of proper output may be used. (See "Power Supply".)

Installation Notes

The installation of this receiver and the elimination of interference are described in the instructions accompanying the receiver. Further information regarding the elimination of interference will be found in Crosley Service Bulletin No. 9.

Circuit

A six-tube superheterodyne circuit is used, employing an oscillating first detector, two intermediate frequency amplifiers, a diode second detector, and two stages of audio frequency amplification. A -39 type tube is used for the first I. F. amplifier; -36 type tubes are used for the first detector and second I. F. amplifier; -37 type tubes are used for the second detector and first A.F. amplifier; and a -41 type tube is used for the output (a -38 type output tube was used in the earlier chasses).

Interstage Coupling.

Air-core transformers are used for coupling the antenna circuit to the first detector and for coupling between all stages except the second detector and first audio, and the first audio and output. Resistance coupling is used in the audio stages.

Tuning Condensers.

The two tuning condensers are ganged and operated simultaneously by the station selector. One section of the gang is shunted across the antenna coupling transformer secondary for tuning to the station frequencies. The other is shunted across the primary of the oscillator coupling coil for tuning to a frequency 181.5 kilocycles higher than the frequency of the antenna coupling circuit.

Power Supply.

The power is supplied by the automobile storage battery and either three or four 45 volt "B" batteries. A "B" eliminator supplying not in excess of 30 ma. at 180 volts may be used. The car storage battery supplies current for the speaker field, for the heaters of the tubes, and for the dial light. The "B" batteries supply current for the plates and screen grids of the tubes.

Fuses.

There is a ¼ ampere fuse in the "B" power cable and a 10 ampere fuse in the "A" power cable.

"A" Circuit.

The pole of the storage battery which is not connected directly to the frame of the car connects through the insulated battery cable lead to the cable connector of the receiver, and thence to one pole of the power switch. (This switch is of the double-pole, single-throw type, one pole making and breaking the "A" circuit, the other making and breaking the "B" circuit.) After going through the switch, the "A" circuit branches, one branch going to the heaters (connected in parallel) of all the tubes and through them to the chassis, another going through the dial light to the chassis, and a third going through the speaker cord, thence through the speaker field to the speaker cable shield, and then through the battery cable shield to the other side of the "A" battery (one connected to car frame).

"B" Circuit

The positive "B" circuit connects through the insulated battery cable lead and cable connector to the "B" pole of the power switch. After passing through the switch, the positive side of the "B" circuit branches, one branch going direct to the screen grid of the output tube, a second branch going through a 25,000 ohm plate coupling resistor to the plate of the first audio tube, a third branch going through the primaries of the interstage transformers to the plates of the I. F. tubes, and a fourth branch going through an 1100 ohm resistor and the oscillator and interstage coils to the plate of the first detector.

The plate of the pentode output tube is connected to the positive side of the "B" circuit through the primary of the speaker input transformer. The screen grids of the first detector and I. F. tubes are connected to the positive "B" circuit through a 20,000 ohm series resistor, and shunted to chassis by a 40,000 ohm resistor. The "B" supply returns to the negative side through the cable shield.

Grid Circuits And Automatic Volume Control.

The output pentode grid is connected to

CROSLEY RADIO CORP.

chassis through a 300,000 ohm grid leak (500,000 ohm in receivers using a -38 type output tube). The level control, a 3 megohm potentiometer, is in the grid circuit of the first audio tube, the grid being connected to the variable contact on the potentiometer. One end of the potentiometer is connected to the second detector coupling condenser, and the other end is connected to the chassis. The grid of the first detector tube is connected directly through the secondary of the coupling transformer to the chassis. The grid circuits of the intermediate frequency stages are connected through isolating condensers to the chassis, and through isolating resistors to the second detector plate circuit. By this means, the plate current flowing in the diode (second detector) load resistors is made to control the bias of the grids of the I. F. tubes, and thereby to automatically control signal levels.

Biasing Resistors

A 750 ohm bias resistor is used between the cathode of the output pentode and chassis. A 2,000 ohm bias resistor is used between the cathode of the first audio tube and chassis. (In chasses using a -38 type output tube the cathode of the output pentode is connected by an 1,100 ohm resistor and a 450 ohm resistor in series to the chassis. The cathode of the first audio tube is connected to the point between the 1,100 ohm and 450 ohm resistors, (thereby receiving its bias through only the 450 ohm resistor).

By Pass Condensers

The 0.25 mf. bypass condensers in the detector plate circuit and a bypass condenser of the same capacity in the screen circuits are combined in one unit. The two I. F. isolating condensers are each of 0.1 mf. capacity, and are also combined in one unit. The audio coupling condenser in the output circuit of the second detector tube is of 0.03 mf. capacity; that in the plate circuit of the first audio tube is of 0.02 mf. capacity. A bypass condenser of 0.0006 to 0.001 mf. capacity is used between the plate of the second detector tube and chassis. The 8 mf. dry electrolytic condenser used from the screen of the pentode to chassis and a condenser of the same value used from the cathode of the pentode to chassis are combined in one unit. A 0.0015 mf. by pass condenser is used between the low side of the oscillator coupling transformer secondary and chassis. A condenser of 0.003 to 0.006 m.f. is shunted from pentode plate to chassis.

Alignment.

The receiver has been aligned at the factory and no realignment should normally be necessary. If it is necessary to realign the receiver, follow procedure given below:

Aligning Intermediate Frequency Stages

The primary and secondary of the transformer between the first detector and the first I. F. amplifier, and the secondary of the transformers between the first I. F. amplifier and the second I. F. amplifier and between the

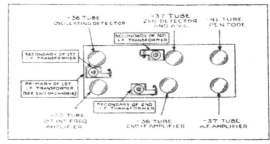

Fig. 1.—Location of Aligning Condensers.

second I. F. amplifier and the second detector, must be tuned accurately to 181.5 kilocycles. (See cut for location).

1. A local oscillator tuned accurately to 181.5 kilocycles is required. Such instruments are supplied by the Weston Electrical Instrument Co., the Jewel Co., the General Radio Co., The Radio Products Co., etc.

2. Set the dial of the station selector to 550 kilocycles.

3. Connect the high side of the test oscillator output through a condenser of approximately 0.1 mf. capacity to the grid of the first detector tube, and the low side of the test oscillator to chassis. Do not remove the clip wire from the grid of the first detector tube.

4. Adjust the two padding condensers at either side of the first intermediate frequency transformer for maximum reading on the output meter.

5. Adjust the secondary padding condensers on the second and third intermediate frequency transformers for maximum reading on the output meter.

When these adjustments have been made, the intermediate frequency stages will be properly aligned.

Aligning Antenna Coupling Circuit and Oscillating Circuit.

The antenna coupling circuit and oscillating circuit should not be aligned until after the intermediate stages have been accurately aligned to 181.5 kilocycles.

CROSLEY RADIO CORP.

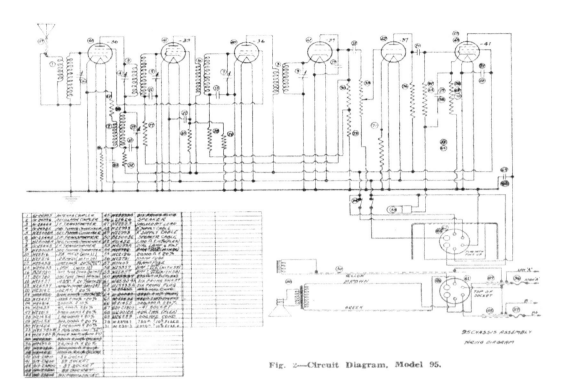

Fig. 2—Circuit Diagram, Model 95.

1. A local oscillator tuned to 1400 kilocycles is required. (As a substitute for the local oscillator, a station of known frequency near 1400 kilocycles may be used, and the dial tuned to the frequency of the station.).

2. Set the dial of the station selector to 1400 kilocycles.

3. Connect the high side of the test oscillator through a 0.00025 mf. condenser (a dummy antenna should be used, if available) to the antenna lead of the receiver,, and connect the low side of the test oscillator to chassis.

4. Adjust the padding condensers on the ganged condenser to give a maximum reading on the output meter.

When these adjustments have been made, the receiver will be properly aligned.

Voltage Limits

The following are the approximate voltages which should be measured at the sockets with tubes in place, speaker connected, and batteries connected. (Four 45-volt 'B' batteries of rated voltage should be used). Check the voltages with a high resistance D. C. voltmeter (600 ohms or more per volt).

Filament Voltage (Measured from high side of filament to chassis)	
All tubes	5.8 to 6.0
Plate Voltages (Measure from plate to chassis)	
1st detector and L. F. tubes	160 to 180
Diode second detector	0
1st A. F. tube	80 to 90
Output tube	150 to 170
Screen Grid Voltages (Measured from screen grid to chassis)	
1st detector and I. F. tubes	70 to 80
Output tubes	160 to 180
Control Grid Voltages (Measured from cathode to chassis)	
1st detector tube	-7 to -9
I. F. tubes	A. V. C. only
Diode (second detector tube)	0
1st A. F. tube	-5 to -6
Output tube	-16 to -18

CROSLEY RADIO CORP.

Model 96 (Roamio)

Installation Notes

Model 96 is a compact automobile receiver. The installation of this receiver and the elimination of interference are described in the instructions accompanying the receiver. Further information regarding the elimination of interference will be found in Crosley Service Bulletin No. 9.

Circuit

This six tube superheterodyne receiver employs a -39 type R.F. tube, a -36 type oscillating first detector, two -39 type I.F. amplifiers, a -85 type diode second detector and A.F. tube, and a -89 type output tube.

Power Supply

The power is supplied by the automobile storage battery and either three or four 45 volt "B" batteries. A "B" eliminator supplying not in excess of 35 ma. at 180 volts may be used. The car storage battery provides current for the speaker field, for the heaters of the tubes, and for the dial light. The "B" batteries supply current for the plates and screen grids of the tubes.

Fuses

There is a ¼ ampere fuse in the "B" power cable, and a 10 ampere fuse in the "A" power cable.

Voltage Limits

The following are the approximate voltages which should be measured at the sockets with the tubes in place, speaker connected, and batteries connected. (Four 45-volt "B" batteries of rated voltage should be used). Check the voltages with a high resistance D. C. voltmeter (600 ohms or more per volt).

Filament Voltages	
All tubes	5.8 to 6.2
Plate Voltages	
R.F., First Det., and I.F. tubes	160 to 200
Second Detector tube	70 to 90
Output tube	150 to 190
Screen Grid Voltages	
R.F., First Det., and I.F. tubes	85 to 105
Output tube	160 to 200
Operating Grid Voltages	
R.F and First I.F. tubes	-3.6 to -4.4
First Detector tube	-6.3 to -7.7
Second I.F. tube	-1.8 to -2.2
Second Detector tube	-5.4 to -6.6
Output tube	-13 to -15

Alignment

The receiver has been aligned at the factory and no realignment should normally be necessary. If it is necessary to realign the receiver, follow the procedure given below:

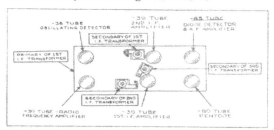

Fig. 1—Location of Aligning Condensers.

Aligning Intermediate Frequency Stages

The primary and secondary of the transformer between the first detector and the first I.F. amplifier, and the secondary of the transformers between the first I.F. amplifier and the second detector, must be tuned accurately to 181.5 kilocycles. (See cut for location).

1. A local oscillator tuned accurately to 181.5 kilocycles is required. Such instruments are supplied by the Weston Electrical Instrument Co., the Jewel Co., the General Radio Co., the Radio Products Co., etc.

2. Set the dial of the station selector to 550 kilocycles.

3. Connect the high side of the test oscillator output through a condenser of approximately 0.1 mf. capacity to the grid of the first detector tube, and the low side of the test oscillator to chassis. **Do not** remove the clip wire from the grid of the first detector tube.

4. Adjust the two padding condensers at either side of the first intermediate frequency transformer for maximum reading on the output meter.

5. Adjust the secondary padding condensers on the second and third intermediate frequency transformers for maximum reading on the output meter.

When these adjustments have been made, the intermediate frequency stages will be properly aligned.

Aligning Radio Frequency Stage And Oscillating Circuit

The antenna coupling circuit, the plate

CROSLEY RADIO CORP

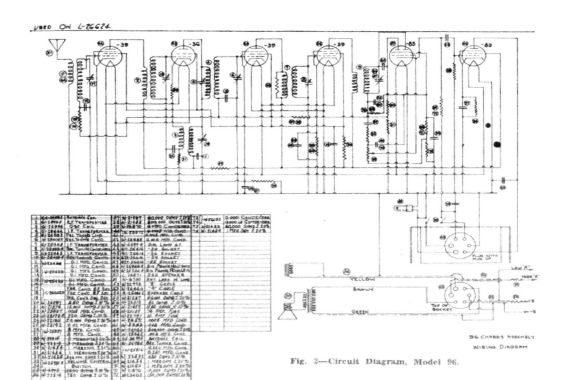

Fig. 2—Circuit Diagram, Model 96.

coupling circuit of the R.F. amplifier, and the oscillating circuit should not be aligned until after the intermediate frequency stages have been accurately aligned to 181.5 kilocycles.

1. A local oscillator tuned to 1400 kilocycles is required. (As a substitute for the local oscillator, a station of known frequency near 1400 kilocycles may be used, and the dial tuned to the frequency of the station).

2. Set the dial of the station selector to 1400 kilocycles.

3. Connect the high side of the test oscillator through a 0.00025 mf. condenser (a dummy antenna should be used if available) to the antenna lead of the receiver, and connect the low side of the test oscillator to chassis.

4. Adjust the padding condensers on the ganged condenser to give a maximum reading on the output meter.

When these adjustments have been made, the receiver will be properly aligned.

CROSLEY RADIO CORP.

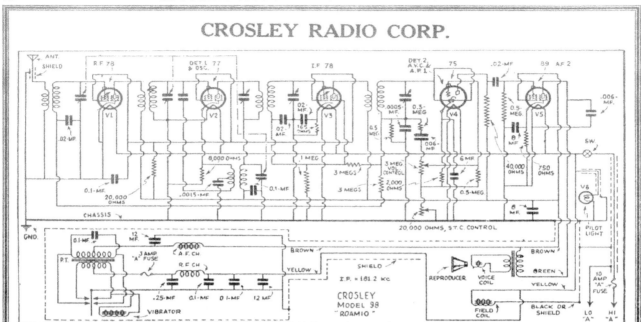

CROSLEY
MODEL 98
"ROAMIO"

Tube	Position	Voltages				
		Plate	Screen Grid	Cathode	Supp. Grid	Fil.
-78	R. F. Amplifier	180	85	0	0	6.0
-77	Oscillating detector	180	85	4.5	4.5	6.0
-78	I. F. Amplifier	180	85	2.0	0	6.0
-75	Diode—A. F. Amplifier	130		1.5		6.0
-89	Output (Class A Pentode)	180	180	17.0	17.0	6.0

"A" battery drain—4.6 amp. at 6.3 volts.

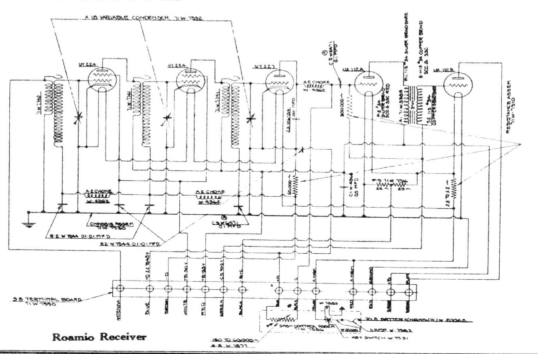

Roamio Receiver

CROSLEY RADIO CORP.

Model 99

Specifications

Model 99 is a six tube superheterodyne designed for automobile operation. The intermediate frequency is 181.5 KC. The "A" supply is furnished by the automobile storage battery and the "B" supply by the automobile storage battery used in connection with a Crosley Syncronode. Service information on the Syncronode is furnished in a separate bulletin.

Tubes and Voltage Limits

The following chart gives the tubes, their functions, and voltages measured with the receiver in operating condition but with no signal to the antenna circuit. Use a high resistance D. C. voltmeter (1000 ohms per volt or more) for all measurements. The voltage limits are + or — 10% of the values given.

All voltages are measured from tube contact to chassis with 6.3 volts at the battery and 170 volts from the Syncronode.

The "Q" control should be entirely off.

| Tube | Position | Plate | Voltages | | | |
			Screen Grid	Cathode	Supp. Grid	Fil.
-78	R. F. Amplifier	170	80	0	0	6.0
-77	Oscillating detector	170	80	4.0	4.0	6.0
-78	I. F. Amplifier	170	80	1.5	1.5	6.0
-85	Diode—A. F. Amplifier	25		2.0		6.0
-89	A. F. Amplifier	170	170	17	17	6.0
-79	Output (Class B)	170		0		6.0

"A" battery drain—5.3 amp. at 6.3 volts.

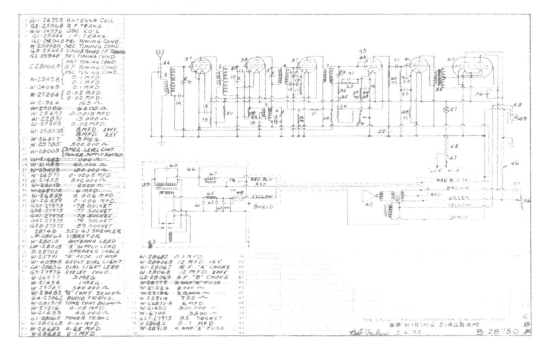

AERIAL DATA

● The following information was compiled by RADIO-CRAFT in response to special requests. This information should be of invaluable assistance to Service Men contemplating the installation of automotive receivers and will be enlarged upon from time to time as additional data are secured.

AUBURN—Closed types on models 8-105 and 12-165 are wired for radio. In Cabriolet and Phaeton sedans, in above models, antenna is optional.

AUSTIN—Radio aerials in Austin cars are not standard but are optional equipment. However, they are factory-installed, under the roof covering of all models, at a slight additional charge.

BUICK—1933 series 50: Business, Sport, Convertible Sedan, and Victoria Coupes are equipped with aerials. 1933 series 60: Sport, Convertible, and Victoria Coupes, Convertible Phaeton, and Sedans are equipped with aerials. 1933 series 80: Sport, Convertible, and Victoria Coupes, Convertible Phaeton, and Sedans are equipped with aerial. 1933 series 90: Sedan model 7-P, Limousine, Club Sedan, Victoria Coupe, and Sedan model 5-P are equipped with aerial.

CADILLAC*—In 1933 closed models: galvanized screen in top; closed models, tinsel tape or braid in top. Lead-in, closed cars, down front; open cars, rear.

CHEVROLET—All 1933 models have aerials built into the headlining. The lead-in is located in the left side front windshield post.

CHRYSLER—Aerials installed at factory in all closed models of 1932 and 1933 Sixes and Eights.

CORD*—No antenna system in 1933 models.

CUNNINGHAM—Antennas installed on order only.

DE SOTO*—1933 closed models: poultry screen in roof; open models, a stranded wire is woven in false headliner on special order only. Lead-in, closed cars, down side; open cars, rear.

DE VAUX*—In 1933 closed models: although there is a poultry screen in roof, an antenna will be furnished on special orders. Lead-in, down front.

DODGE—1933, 8 cylinder: 4-door Sedan, 5-passenger Coupe, Rumble Seat Coupe are equipped with aerial. 6 cylinder: 4-door Sedan, 2-door Sedan, Business Coupe, Rumble Seat Coupe are equipped with aerials. An opening in the floor is provided for the installation of a radio set manufactured for Dodge cars by Philco Transitone.

DURANT*—No antenna system in 1933 models.

DUESENBERG—Aerials installed on special order only.

ESSEX**—1933 models: stranded wire antenna in top only on special order.

FORD—The following 1933 cars are installed with an aerial: Tudor, Fordor, and Victoria. The following 1933 models have running board aerials: Roadster, Phaeton, Cabriolet, 3-window Coupe, and 5-window Coupe.

FRANKLIN—1932-1933 series 16 and 17 have insulated chicken wire in roof which may be used as an antenna. The series 18, of the same years, have grounded chicken wire.

GRAHAM-PAIGE—Aerials installed on special order only.

HUDSON—1933 Terraplane "6" and "8" and Hudson "6" and "8": all closed cars equipped with antennas.

HUPMOBILE—1933 models 321, 322, and 326 are equipped with antennas; antenna lead-in under cowl.

LA SALLE*—1933 models: galvanized screen in top; open models, tinsel tape or braid in top. Lead-in, closed cars, down front; open cars, rear.

LINCOLN—All closed models are equipped with antennas.

MARMON—Standard equipment in 1930 models 69, 79, and Big Eight; 1931, models 88. The 1932 and 1933 16-cylinder cars are not equipped with an aerial but there is a wire mesh, in the head-lining, to which a lead-in may be attached and run down the left side front windshield post.

NASH—1932-1933, series 1000: first batch of cars equipped with antenna, with no lead-in. Later models, lead-in down left center post; present model, lead-in down right front corner post. Sixteen spark plug suppressors and two distributor suppressors, standard equipment in twin-ignition models, series 1080, 1090, 1180 and 1190.

OLDSMOBILE—1933, 6 cylinder models: Business, Sport, 5-passenger, Touring Coupes, 4-door Sedan, 4-door Touring Sedan, and 5-passenger models are equipped with aerials. 1933, 8 cylinder models: Business, Sport, 5-passenger Coupes, 5-passenger Touring Coupe, 4-door Sedan, and 4-door Touring Sedan are equipped with aerials.

PACKARD—1933 models: All closed cars equipped with aerials.

PEERLESS*—1933 closed models: poultry screen in roof. Shielded lead-in, closed cars, down front.

PIERCE-ARROW—1930 models A, B, and C equipped with aerials. 1931 models 41, 42, and 43 equipped with aerials. 1932 models 51, 52, 53, and 54 equipped with aerials. 1933 models 836, 1236, 1242, and 1247 equipped with aerials.

PLYMOUTH—Antennas are factory-installed in 1933 models.

PONTIAC—1933, 8 cylinder models are equipped with roof antennas with the lead-in behind the right cowl trim pad. No other models are equipped with aerials.

REO*—Flying Cloud model 52: equipped with antenna, only on special order. Royale N-2 for 1933, equipped with antenna. Open and convertible Reo's will be supplied with antenna system on order.

ROCKNE*—1932 models: Sedan and Coupe are equipped with aerials. The Convertible Sedan and Convertible Roadster are not equipped with aerials. 1933 models: Sedan and Coupe are equipped with aerials. The Convertible Sedan and Convertible Roadster are not equipped with aerials.

ROLLS ROYCE*—Special bodies. Antenna can be furnished.

STUDEBAKER—The following Studebaker cars are equipped with aerials: 1930 starting July, models 70, 80, and 90; starting June, 1931, models 61 and 54; 1932 models, 91, 71, 62, and 55; 1933 models, 56, 73, 82 and 92. In 1933 Convertible Sedans and Roadsters aerial is installed at factory as optional equipment, at slight extra cost.

STUTZ—Aerials installed on special order only.

TERRAPLANE**—1933 models: stranded wire antenna in top, only on special order.

WILLYS KNIGHT*—1933 closed models: stranded wire in top; open models, none. Lead-in, closed cars, down front.

WILLYS OVERLAND—1933 model 99 is equipped with aerial.

———

(*Courtesy; Philco Transitone. **Courtesy, Zenith Radio Corp.)

DELCO APPLIANCE CORP.

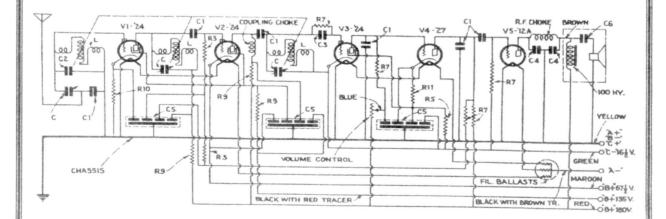

FADA RADIO & ELECTRIC CORP.

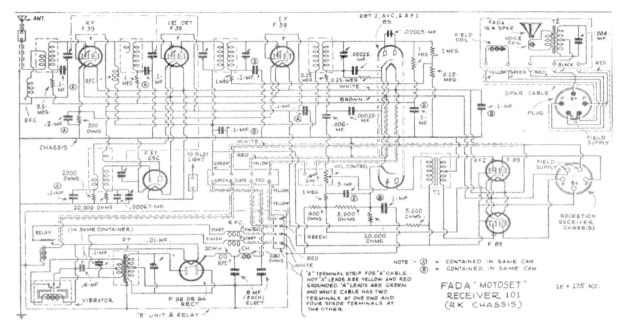

FADA "MOTOSET"
RECEIVER 101
(RK CHASSIS)

I.F. = 175 Kc.

COMPRISES three units: a control head that clamps to the steering column, a fully enclosed receiver unit including plate supply device, and a separate dynamic loudspeaker. A flexible shaft connects the first two units. The receiver box measures $10\frac{1}{8}$ inches long, $7\frac{1}{8}$ inches high and $7\frac{3}{4}$ inches deep; the speaker, circular in shape, is $9\frac{1}{2}$ inches in diameter, $4\frac{1}{2}$ inches deep. Noise suppression equipment is furnished. Automatic lock closes when volume control is turned to minimum setting; a key releases it. Antenna not included.

Backlash in the remote control has been eliminated by a worm gear drive on the condenser shaft, with constant tension maintained by a strong spring.

Eight tube superheterodyne circuit, with automatic volume control. Tubes used: three type 39, one 37, one 85, two 89, one 98. Current drain from car's storage battery, $6\frac{1}{4}$ amperes. I. F., 175 kc. Sensitivity 1 m.p.m. Output 3 watts. Plate power unit is of the vibrator type with a full-wave mercury-vapor rectifier tube. Mechanical hum eliminated by sealed lead housing, $\frac{3}{8}$ inch thick.

An automatic relay is incorporated in the chassis to prevent operation of the set on a low storage battery.

P. R. MALLORY, INC.

MALLORY ELKONODE

THE Mallory self-rectifying Elkonode is a device which, within itself, sets up the essentially alternating current and rectifies it at the same time, thus eliminating the cost of the rectifying tube and other parts usually required with the more conventional D.C. to A.C. converters. The construction is such that it permits a design which is silent and which greatly reduces the over-all dimensions of the unit. A complete filter system is incorporated, as shown in the schematic diagram here. There are five different types of Elkonodes in different ratings:

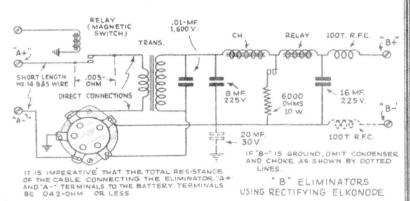

IT IS IMPERATIVE THAT THE TOTAL RESISTANCE OF THE CABLE CONNECTING THE ELIMINATOR "A+" AND "A-" TERMINALS TO THE BATTERY TERMINALS BE .042-OHM OR LESS.

IF "B-" IS GROUND, OMIT CONDENSER AND CHOKE AS SHOWN BY DOTTED LINES.

"B" ELIMINATORS
USING RECTIFYING ELKONODE

Schematic circuit of the Elkonode described in the text. This unit is self-rectifying—no rectifier tube is used. This method accounts for the high efficiency of this unit.

FORD-MAJESTIC

FORD-MAJESTIC MODEL 114 AUTO-RADIO SET

"A FLOOR board installation," characterizes the Ford-Majestic 6-tube car radio receiver. The set chassis and the motor generator drop into water-tight, drawn-sheet-steel pans recessed in the floor; both containers measure 6-1/32 x 9-7/32 long, inside dimensions. The arrangement is illustrated.

A complete Ford-Majestic installation includes the receiver chassis and tubes, 180 volt motor-generator "B" supply, steering-column control head, separate 6-in. dynamic reproducer, generator and distributor filter condensers, spark plug suppressors, fuses, running-board antenna, necessary shielded power and control cables, and shielding and miscellaneous hardware. The output transformer is incorporated in the set.

There is no "fiddling" with leads; polarized-plug connections are provided on the four major units: set, "B" supply, reproducer and control head. As shown in one of the illustrations, the "B" and set units mount just in front of the rear seat in pans recessed into the floor-boards. (Complete servicing may be accomplished inside the car, within one hour.) The control head is designed for adjustment by an operator wearing gloves. The reproducer, which acts as a terminal board for the set and control cables, is supported on a bracket which mounts under the cowl directly above the steering column—this permits the car heater to be retained.

A superheterodyne circuit is used; the I.F. is 175 kc. The chassis is the Majestic model 114; the diagram is shown here. Tubes used: one G-39, first R.F.; one G-38, first detector and oscillator; one G-39, first I.F.; one G-85, second-detector and first A.F.; two G-38's, push-pull second A.F. The total "A" drain is 5 amperes; "B", 36 milliamperes supplied by a motor-generator. This set has a sensitivity minimum of 3 microvolts and selectivity of 30 kc. at 10,000 times; maximum noise level (antenna disconnected), 1 volt.

The reproducer is not designed for very low-note reproduction, since the car noises completely obliterate reproduction in the lower register. Half-wave detection is obtained in the 85. Automatic volume control and silent tuning control are features of the Ford-Majestic Model 114 automotive radio set. The cable shielding functions as the ground, common return for the "A" and "B" voltage, and reproducer voice currents.

Operating voltages of this receiver:

Tube	Plate Voltage	Screen Voltage	Cathode Voltage
V1	180	85	0
V2	180	85	15
V3	180	85	1.1
V4	A.F. Plate 50		2
V5	170	180	17
V6	170	180	17

Note: Measurements made with 1000-ohm-per-volt meter with 300 volt range, all tubes in their sockets and receiver connected to a storage battery supply delivering 6 volts at the cable terminals, under load.

Readings to be taken from designated points to ground.

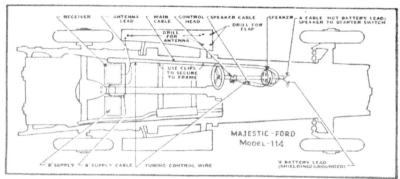

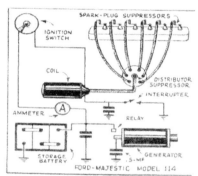

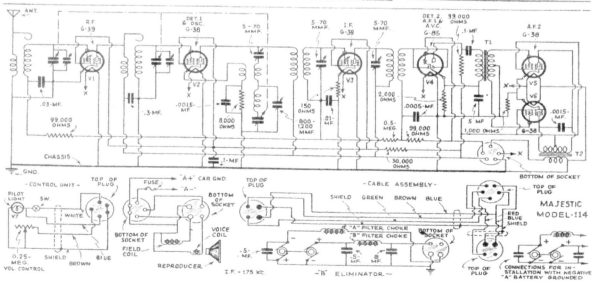

FRANKLIN RADIO CORP.

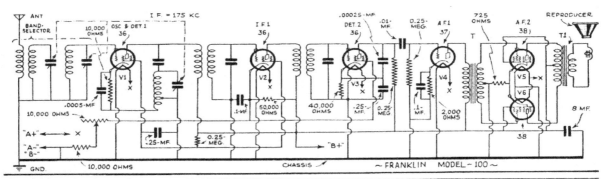

~ FRANKLIN MODEL - 100 ~

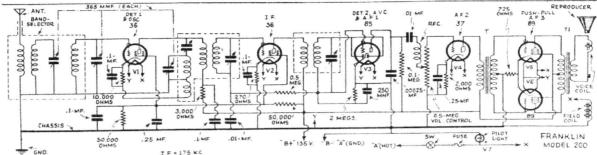

FRANKLIN MODEL 200

FRANKLIN MODEL 100

THE Franklin Model 100 auto-radio receiver is a six tube set using a dynamic speaker, equipped with an under-car antenna and all accessories required to operate the set. The accessories include all spark-plug and distributor suppressors, generator condenser, antenna, loudspeaker of the dynamic type, batteries, battery box,

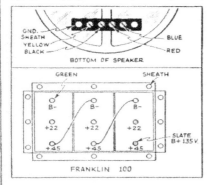

FRANKLIN 100

tubes, remote-control tuning unit, and brackets for same.

The receiver is designed for an under-car antenna which preferably should be placed under the car as low as possible and extending from the front to the rear axle. The receiver proper is so designed as to mount either on the steering column (as shown in the illustration) or on the dash in either the driver or engine side of the bulkhead. If possible, the set should be mounted on the right-hand side of the driver's compartment side of the dashboard when not mounted on the steering column.

The "B" battery box should be mounted on the floor boards, if possible. When mounting the battery box, be sure that the location is such as not to interfere with the normal operation of brake rods, etc.

The circuit reproduced here, is of the superheterodyne type and consists of a combination detector-oscillator tube with a band-pass input stage using a 36 tube; an I.F. amplifier using a 36 tube and tuned to 175 kc.; a second detector using a 36; a first audio

stage using a 37; and a push-pull output stage incorporating two type 38 pentodes in push pull. When first operated, the antenna stage must be resonated to the particular antenna used, by rotating a small screw located on the left hand side of the set.

The "B" source consists of three 45-volt batteries housed in a box, as previously described. The color code of the wiring to both the speaker and the batteries are illustrated here.

FRANKLIN MODEL 200

THIS is a six-tube superheterodyne, with steering column remote control. The receiver, minus plate supply, and the loudspeaker are separate units. No plate power device is furnished, but a metal container for "B" batteries is included, as are ignition suppressors. The set may be mounted on the steering column or under the dashboard. The speaker is of the dynamic type.

Tubes used are two 36's, one 37, one 85 and two 89's. Requires 135 volts of "B" from either batteries or any auto-radio power unit of proper capacity.

CHAS. HOODWIN CO.

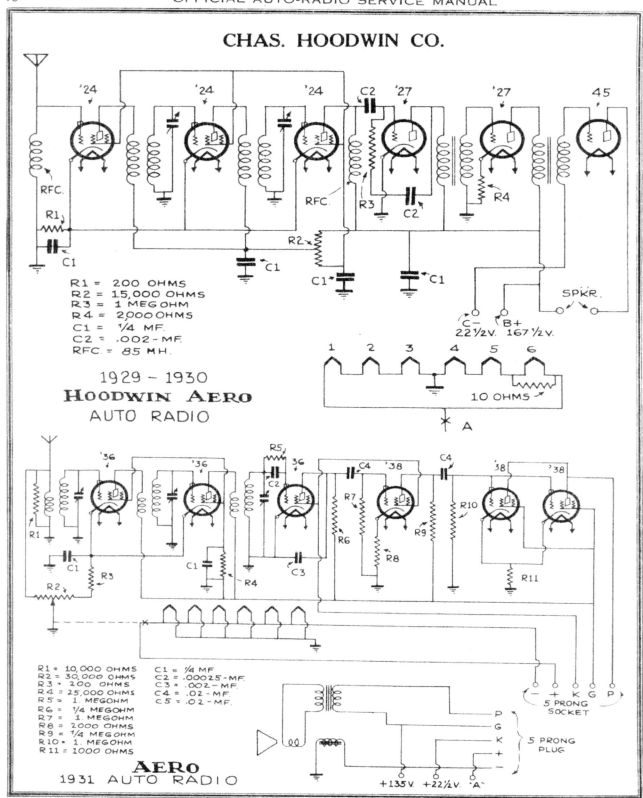

R1 = 200 OHMS
R2 = 15,000 OHMS
R3 = 1 MEG OHM
R4 = 2000 OHMS
C1 = ¼ MF.
C2 = .002 -MF.
RFC. = 85 M.H.

1929 – 1930
HOODWIN AERO
AUTO RADIO

SPKR.

C– B+
22½V. 167½V.

10 OHMS

A

R1 = 10,000 OHMS
R2 = 30,000 OHMS
R3 = 200 OHMS
R4 = 25,000 OHMS
R5 = 1. MEGOHM
R6 = ¼ MEGOHM
R7 = 1. MEGOHM
R8 = 2000 OHMS
R9 = ¼ MEGOHM
R10 = 1. MEGOHM
R11 = 1000 OHMS

C1 = ¼ MF
C2 = .00025 -MF.
C3 = .002 -MF.
C4 = .02 -MF.
C5 = .02 -MF.

5 PRONG SOCKET
– + K G P
5 PRONG PLUG
P G K + –
+135V. +22½V. 'A'

AERO
1931 AUTO RADIO

FEDERATED PURCHASER, INC.

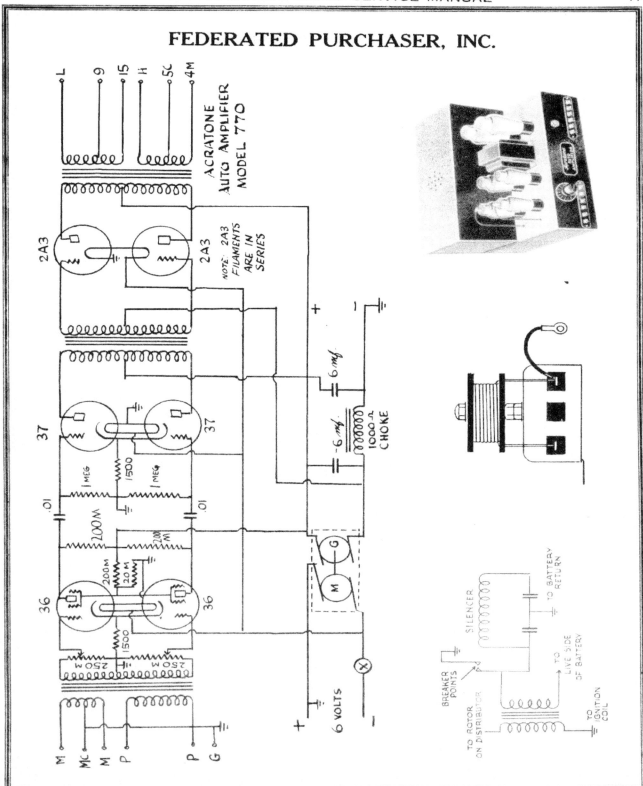

ACRATONE
AUTO AMPLIFIER
MODEL 770

NOTE: 2A3 FILAMENTS ARE IN SERIES

2A3   2A3

37   37

36   36

1 MEG   1500   1 MEG

.01   .01

200M   200M

200M   20M

1500

250M   250M

6 mf.

.6 mf.   1000-Ω CHOKE

6 VOLTS

SILENCER

TO BATTERY RETURN

TO LIVE SIDE OF BATTERY

TO IGNITION COIL

BREAKER POINTS

TO ROTOR ON DISTRIBUTER

GALVIN MFG. CORP.

MOTOROLA SERVICE MANUAL

CHAPTER I

INTRODUCTION:

The installation of Motorola Auto-Radio is comparatively simple. It is merely a matter of understanding and taking pains, together with observing and carrying out essential details. While Motorola design effectively meets the requirements necessary for simple installation, the problems encountered in various makes of automobiles presents ample opportunity to exercise skillful ingenuity which makes installation interesting and enjoyable.

Motorola is the direct result of experience and specializing in automobile radio. It is entirely a different problem than home radio design. There are definite reasons for doing certain things. In automobiles, various types of interference, acoustics and car design must be considered.

An auto-radio to give customer satisfaction should equal home performance. It should be extremely sensitive. It should not be too selective, since hair-line selectivity is a distinct disadvantage in auto-radio. It should be conservative in battery consumption.

The station selector should be visible by the driver from his normal position in the seat. There should be no back lash in the tuning device. The operation of volume control should not affect tone quality. The speaker should be flexible enough to take full advantage of cowl acoustics. There must be no fading out of stations --- the receiver must maintain steady, even volume, through efficiently designed Automatic Volume Control. The set should be simple and easy to install. And it should be possible to mount the set on either engine or driver's side of the bulkhead. Most important, the set should be thoroughly shielded.

Check Motorola Auto-Radio and see how completely every essential has been covered. Motorola was the first to embody in car radio such outstanding features as Steering Post Control, Dynamic Speaker, Push-Pull, Pentode Tube, Perfected A.V.C., and All-Electric self-contained receiver. It is designed to give customer satisfaction -- and profit to the dealer and installation man. While simplicity is the keynote, always bear in mind that the performance of Motorola is no better than its installation.

Motorola installation has been made simple. Follow the instructions closely, which are packed with each receiver --- use your ingenuity --- take pains --- and you will profit not only in dollars and cents, but in the knowledge that you have given customer satisfaction. And satisfied customers mean continued profits for you.

Before you make an installation read the instructions carefully. Then read them again. Keep them handy for reference from time to time --- and observe every detail.

In approaching the installation problem, there are essential fundamentals vital to the performance of Motorola which you should be familiar with --- and thoroughly understand.

INTERFERENCE:

The interference angle must be taken into consideration. There are two classes of interference encountered in auto-radio, chassis pick-up and antenna pick-up. Chassis pick-up, which is a motor interference heard from the loud speaker with the antenna of the radio disconnected, is the most difficult of all to defeat. It is a direct indication that the waves set up by the car are passing through the radio and are being amplified as radio frequency, thereby, causing an effective disturbance in the loud speaker. Motorola engineers have combatted the trouble effectively by the use of thorough shielding, the wiping of all contact surfaces, the double shielding of all R.F. coils and perfected cable shielding.

An installation man must understand that an automobile has a series of currents flowing through the entire body, bulkhead or chassis, and that an auto-radio installed in an automobile comes in contact with nearly all parts of the automobile, the bulkhead, channel frame, wind shield and storage battery. The currents flowing through them are excited by the return circuit of the high voltage from the ignition coil and unless the radio is thoroughly protected against these currents the installer will run up against an almost hopeless job of defeating motor noise.

Bonding and coupling the various body parts of the car so as to allow the high voltage from the secondary of the ignition coil to return to the motor block in but one path has been the most common method used to eliminate motor noise in the past. However, the design of Motorola eliminates, in nearly all cases, the necessity for thorough bonding. It is necessary only as an expedient to suppress another type of interference known as "antenna pick-up" which is generally due to the radio frequency currents generated by the ignition system which leak out from the hood. A few simple procedures, usually take care of the majority of installations in this respect.

GALVIN MFG. CORP.

The convenience of having the speaker a separate unit should be clear to the service man. Our speaker is housed in a small, rounded case with a single stud mounting connected to a very effectively shielded cable, thoroughly by-passed with filter condensers or shielded to reduce static charges to a minimum.

While the above suggestions are very helpful in getting practically perfect reproduction, they must not be misconstrued as meaning that the location of the speaker is extremely critical for good reproduction. In order to satisfy a very critical musical ear, however, it will be extremely helpful to follow the suggestions outlined.

SET LOCATION:

The following is a suggested location of auto-radios in automobiles that we have found to be what we class as a logical place in which to install a radio. If the auto-radio is placed in the suggested location you will find that the installation will be the most economical and also the quickest way to install the radio.

There may be special cases wherein the owner has something installed in that particular location and it will be found easier to install the radio at a different location than to remove the appliance.

The list is applicable to cars of the 1931 series. In mounting any of the auto-radios care should be exercised in spacing the can away from the bulkhead by means of the nuts and washers furnished. This avoids clamping the cables or any of the wires that happen to protrude through the bulkhead behind the radio.

LOCATION OF RECEIVER:

Ford, Model "A" ... Motor compartment on the left side.

Chevrolet ... Motor compartment on the left side or below the cowl on the right inside. (To install 1932 model set remove carburetor and air cleaner temporarily).

Buick ... Below the cowl on the right inside.

Chrysler ... Right side of car under cowl.

Pontiac ... Left side of motor compartment or right side of car under cowl.

DeSoto ... Left side of motor compartment or right side of car under cowl.

Plymouth ... Left motor compartment or right side of car under cowl.

Cadillac ... Left motor compartment or right side of car under cowl.

Lincoln ... Center of motor compartment or right side of car under cowl.

Packard ... Center motor compartment or right side of car under cowl. (Light Eight - Left side motor compartment).

Oakland ... V8 .. Below cowl on right side or right side of car under cowl.

Studebaker ... Left motor compartment or right side of car under cowl.

Oldsmobile ... Right motor compartment or right side of car under cowl.

Auburn ... Right motor compartment or right side of car under cowl.

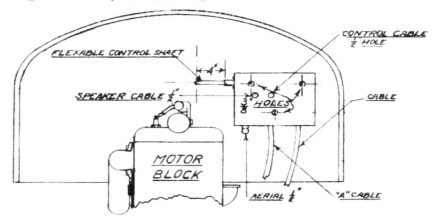

Figure 2

GALVIN MFG. CORP.

As previously outlined, due to the thorough shielding of Motorola auto-radios, all models can be mounted on either side of the bulkhead. The bulkhead being that partition in the car which separates the motor compartment from the driver's compartment. If the auto-radio is mounted in the motor compartment keep in mind the long easy curve of the control shaft running from the set to the control panel. It should not be less than four inches (4"), as shown in Figure 2, from the hole in the bulkhead to the side of the radio.

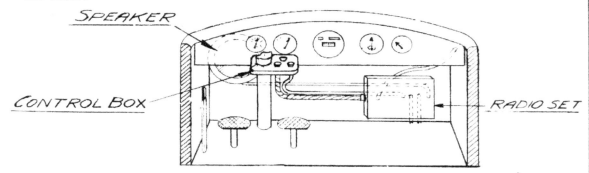

SPEAKER

CONTROL BOX

RADIO SET

Figure 3

Figure (3) illustrates a typical driver's compartment installation, and with such an installation it will not be necessary to disconnect wires from any unit, merely drill four (4) small holes in the bulkhead properly spaced and located by means of a template --- furnished with the package. After making proper electrical power connections in accordance with the tags, as located on each wire, the radio will function.

These cables, after being connected, should be bracketed or fastened in place by means of the small clips furnished with each radio, or -- if preferred -- the installer may tape them. If he tapes them, however, care must be exercised to avoid pulling the cables up too tight as this is detrimental.

ANTENNAS:

There are various ways to obtain energy or antenna signals. Different makes and types of cars have various conditions and each must be coped with individually. Our experience has shown that the roof antenna, if properly installed, is the most satisfactory.

The most satisfactory roof antenna is a piece of copper or galvanized screen, approximately 3 feet square installed between the head-lining and roof of the car. This is done by dropping the head-lining back for a distance of one yard or more and tacking the screen to the ribs. The screen should not come closer than 8 inches to the metal on top at the front of the car and to within 4 inches of the metal on the sides of the top.

If after dropping the head-lining it is discovered that chicken wire is used in the construction of the top, it will not be satisfactory to install the screen as described in the above paragraph. Instead check the chicken wire with a continuity meter to see if it is grounded. If it is not, a lead may be attached and the chicken wire used as an aerial. If it proves to be grounded it must be freed in the manner described in a later paragraph on "Roof Antenna in Model A Fords".
The following automobile manufacturers announce roof antenna in various 1932 models:

TYPE AUTOMOBILE	YEAR MODEL	REMARKS
Chrysler	1932	Roof antenna with lead-in and provisions for "B" Battery Box
Dodge	1932	"
DeSoto	1932	"
Plymouth	1932	"
Reo	1932	Equipped with roof antenna and lead-in
Rockne	1932	"
Studebaker	1932	"
Buick	All Models	$6.00 additional for antenna installation
Franklin	1932	Roof antenna, no lead-in
Cunningham	All Models	additional charge for antenna installation
Ford	1932	Roof antenna, but no lead-in.

On the balance of the automobiles it will be necessary to install an antenna on the car.

CHECK THE ANTENNA:

The antennas that have been installed by the manufacturers will, for some time to come, need to be checked very thoroughly. It has been recently found that many installation stations, in installing auto-radio, fail to make a thorough check of antenna installations after installation or have failed to check it if the antenna installation had been put in by the automobile manufacturer. It can be easily checked by simply trying

GALVIN MFG. CORP.

to peak the antenna stage.   If you are unable to reach a peak on the antenna assembly you have either a bad, leaky antenna, or one with too great capacity. We quote herewith a section from our Service Instruction Sheet which deals with this phase:

"After the set is installed ready for operation, it may be necessary to balance the set with the antenna.   This is done by adjustment of the first antenna trimmer.   Openings for this adjustment are provided for in the various models."

In making this adjustment be absolutely sure that you have properly tuned in a very weak station around 20 or 30 on the dial, adjust the trimmer in and out with a screw driver until the point of maximum volume is reached.   DO NOT DISTURB THE TRIMMER BELOW THE OTHER TWO SMALLER HOLES AT THIS POINT.   These have been very carefully adjusted at the factory.

You can check whether the antenna is grounded by means of a very sensitive voltmeter, such as 200-volt, 1000-ohm per volt voltmeter placed in series with 200-volts of "B" Battery, touching one end of the meter to the antenna and the other end of the batteries to the chassis of the car. With this sensitive meter and this high voltage, you should not get over a 2-volt deflection on the meter, even on a damp day. If you do get over a 2-volt deflection it indicates that the antenna is either fully or partially grounded, depending on reading. If a reading is obtained it will be necessary to remove the head-lining and cut a strip three or four inches wide out of the screen wire or around its edge, thereby insulating and isolating it from the frame of the car.   If a dome light is installed in the car, a circle should be cut out of the screen so it will not be near the dome light.

An effective area of this screen need not be greater than 9 square feet.   Bearing this in mind, you will not find it necessary to take the head-lining down all the way back. Generally to the second rib is sufficient. If, after freeing the screen from the end supports, it is detected that there is a chance of the screen shifting, tacking the screen to one of the ribs will hold it in place.

The lead-in for any of the above type of installations, must be given consideration and it should be brought down on the same side of the car where the Radio is mounted and down the front corner post, either right or left, depending of course on the position of the Radio.   On many cars, you will find the windlass is composed of a hollow rubber tube and makes a very nice housing for the lead-in wire and having a distinct electrical advantage insofar as it keeps the wire away from the metal of the car, maintaining the capacity of the lead-in very low.

PLATE ANTENNA:

If it is desired, a plate antenna may be used. The plate consists of a piece of metal, approximately 2½ square feet in area, rigidly held to the car and the closer to the ground this is placed, the greater efficiency within of course practical limits. It may be placed under the running boards or fastened to the channel frame.   These plates may be obtained from Galvin Manufacturing Corporation on special order, and are fastened by means of clamps to the frame of the car, no drilling being necessary.

Most serious consideration in the use of the plate antenna is the lead-in from the plate to the Radio set.   The lead-in must be of sufficiently insulated wire so in fastening it rigidly to the frame, the capacity of that wire with respect to the frame is small, also the wires must be fastened so that they cannot be easily knocked off.

In the use of a plate or under-car aerial, some additional shielding may be needed on the antenna lead. If the unshielded portion of the antenna lead is over one foot in length a piece of loom, similar to that used on the shielded part of the lead, should be used to keep the shielding from coming too close to the antenna lead wire.   Enough of this loom should be slipped over the wire to reach within about four inches of where the lead attaches to the aerial proper.   Braided sheathing is then slipped over this loom, and joined to the shielding of the shielded lead from the set so as to make a continuous shielded lead from the set to within about four inches of aerial proper. The end of shield nearest the aerial should then be grounded to frame of car.

UNDER-CAR ANTENNA:

The Under-Car antenna consists of a wire fastened from the lower point on the right hand side of the rear axle to the lowest point under the motor, then back to the lowest point on the left hand side of the rear axle, thus forming a "V". At the vertex of the "V" a heavy coil spring should be attached to keep up slack, the spring being insulated from the motor, as well as the other two ends of the wire.   The lead-in, of course, is fastened at the vertex.

ROOF ANTENNA ON MODEL "A" FORDS·

In the application of the roof antenna on the Model "A" 1930 Fords, when the top is dropped you will notice that No. 2 rib is a steel rib and it will be necessary in order to get full effect from the antenna, that the screen be cut clear of this steel rib as shown in Figure (4).   The Figure (4) should clearly illustrate just what is meant and it retains the excellent antenna characteristics that could be obtained from the older model Fords.

GALVIN MFG. CORP.

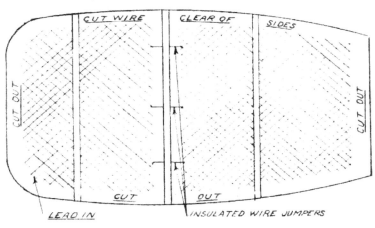

Figure 4

GENERAL IGNITION AND GENERATOR SUPPRESSION:

Motorola receivers are very sensitive. In fact, more so than the average home-radio receiver. Therefore, it is sensitive to every weak impulse such as static and ignition or generator interference. It will be necessary, generally, to use suppressors with the auto-radio and to prevent re-radiation it will be necessary to even go farther than the installation of suppressors.

Securely mount the spark plug suppressors to each spark plug and connect the spark plug wires to the other end of the spark plug suppressors. Also mount and connect the distributor suppressor at the DISTRIBUTOR. Suppressors are nothing more or less than carbon resistors, having approximately 25,000 ohm resistance and their effect or purpose is to reduce the surge of the high voltage impulse. They in no way interfere with the running of the engine, provided the spark plug points are opened approximately 3/64 of an inch. These resistors are tested at the factory and should be re-checked by you after installation by means of a continuity tester.

You will also find in the package a condenser with a pig-tail lead. Apply this condenser at the generator, fasten the condenser in position by removing a screw from the frame of the generator or thereabouts, putting it through the hold on the clamp of this condenser. Fasten the pig-tail lead to the battery side of the cutout switch. If substitution of this condenser is necessary at any future time, care must be taken that it is a non-inductive condenser of not less than ½ mfd... preferably a 1 mfd.

As previously outlined, chassis pick-up with Motorola should not be expected but it is occasionally run into when conditions about the car are subnormal ... such as the ignition coil arcing internally. This should be eliminated by substitution of another coil. Other causes may be ... a run-down storage battery, a defective storage battery connection, a defective ground connection,. or the battery ignition points or distributor points in bad order. Any, or all, of which must be checked in case interference is received with antenna disconnected.

Satisfactory operation, however, will never be received until this class of interference is absolutely eliminated. It has occasionally been found where the installation man has not connected the "A" battery leads in accordance with the directions on the wire, having been placed at some other place other than specified. Or sometimes the sheathing has been torn off the cable, therefore rendering the shielding useless. The shielding most commonly being torn by pulling the cable through too small a hole in the bulkhead. In this case, slide an additional piece of copper sheathing over torn piece. If you have no chassis interference then connect the antenna lead and if interference is then detected it is all being received from the antenna. The hood should be put in place and the catch fastened. If no difference is then noticed you are receiving your pick-up from the dome light, from a leaky hood, or from the high tension coil circuit, electric wind shield wiper, cigarette lighter, or some other wire that is feeding the energy from the motor compartment to the antenna proper. Sometimes the hood is not grounded. It is then necessary to scrape point off the sides of the hood supports.

The interference caused by any one of the aforementioned wires can be eliminated by the use of an ingenuous dome light filter. The application of the dome light filter used in filtering the dome light wire is shown in Figure 5.

This filter can be obtained from any Motorola Distributor. It is wound with a heavy gauge wire, allowing it to be used in a number of places without causing any appreciable drop in voltage.

GALVIN MFG. CORP.

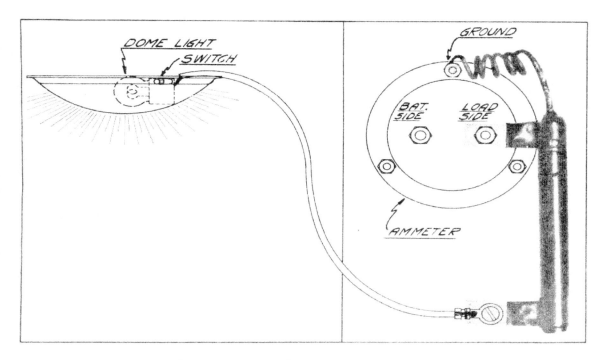

BACK VIEW OF INSTRUMENT PANEL.

Figure 5

Another quite common cause of feeding the energy to the antenna circuit is caused by the ignition coil being located on the instrument board. In this case it will be necessary that you apply shielding over the cable grounding this shield where it passes through the bulkhead, on the motor compartment side – being careful to keep shielding spaced at least 1½" from coil and suppressor. If unglazed rubber covered high tension cable is in use on car, this will need to be replaced with Packard Hi-Tension cable before shielding can be installed. You will find it unnecessary to shield any more of the high tension wires than this one main distributor wire. We have found it unnecessary in any case to shield the entire ignition system.

But if you have noticed a distinct change in shielding this wire but the interference is not entirely eliminated, then the point at which this wire passes through the bulkhead should be moved. That is, a separate hole of its own should be drilled in the bulkhead and this wire should be passed through it so that the cable, in its path from the ignition coil to the distributor, will be as short as possible in the driver's compartment, keeping it free from all other wires connected to the "A" Battery circuit.

There have been a few cases where reversing the connection of the ignition coil has helped as there are times when these connections have been reversed during the production of the automobile.

****

There are a few things in the suppression of antenna pick-up in various makes of cars which might be worthy of mention.

GALVIN MFG. CORP.

AUBURN:

The majority of Auburns will be found to operate very satisfactorily on one suppressor, that being applied in the line between the ignition coil and the distributor.

The aluminum plate which houses the distributor must be thoroughly grounded, both top and bottom, and is most easily accomplished by riveting a piece of shielding braid on to the cover under the aluminum cover and carrying this shield down, fastening it under one of the head bolts.

Then remove the black and yellow wire on the ignition coil .. the other end of this wire is at the electrolux switch ... and replace this wire with a shielded wire, grounding this shielded wire where it passes through the bulkhead.

This should take care of the 1930 and 1931 Auburns.

BUICKS: (1929—30—31)

Due to the spark wires all being thoroughly shielded, the application of one suppressor is all that is necessary on a Buick. This suppressor should be applied as close to the distributor as it is possible to make it as the antenna pick-up is very severe. Grounding the wind shield, as well as the small metal pieces on both sides of the wind shield, will be found very effective, when a roof aerial is used as there are a number of Buick models that do not have these parts grounded.

CHEVROLET:

If the car is not a new model contact points should be examined thoroughly and if any of them have been pitted new contact points should be installed.

Apply an extra condenser at the ammeter, dome light filter in the dome light circuit if connected, and with a short piece of shielding bond the rain spouting which is the small angular material running close around the edge of the car roof. This has been discovered not to be grounded in the majority of Chevrolets and it will be necessary, after bonding it together, to then ground it to a corner post, checking thoroughly to see that the corner post used is likewise grounded.

Then abide with the same type of interference elimination used in the Buick which will effectively take care of this car.

It has also been found in some cases in this car that the person sitting on passenger side will radiate interference carried from his feet which are close to coil on other side of bulkhead, up to the antenna. A piece of screening placed under the floor mat will eliminate this type of interference. This screen must be grounded.

DODGE:

It is necessary that there be thorough shielding of the cable leading from the ignition coil to the bulkhead, grounding the shield to the outside of the bulkhead. An additional heavy bond must be made from the motor to the bulkhead, in some cars, or from the motor to the channel frame in others.

ESSEX:

It is very important in the Essex that the "A" Battery connections be made to the storage battery. It will also be necessary in all installations to install a by-pass condenser at the ignition switch. This condenser should be at least 1 mfd.

FORD: Model "A"

It will always be necessary in Fords to bond the spark control rod to the motor by means of a piece of shielding, soldering one end of the shielding to the rod and the other end under a cylinder head bolt.

It has occasionally been necessary to place an additional bond to the other end of the spark control rod to the bulkhead.

In a few instances it has been necessary to bond the electrolux cable to the bulkhead at the point where it enters the small rubber terminal block.

The distributor spacing must be checked up thoroughly to see that is is not too large, as this varies considerably in Fords. If it is found to be over five thousandths of an inch it should be built up with solder or piened. Figure #6 indicates what is meant by building up the distributor.

GRAHAM-PAIGE:

The shielding of the wire from the ignition coil to the switch located on the steering column, grounding this shield to the bulkhead, is necessary in the Graham-Paige. It is also necessary to place an additional by-pass condenser at the fuse block, located on the bulkhead, together with the standard suppressors. This will take care of the majority of this type car.

GALVIN MFG. CORP.

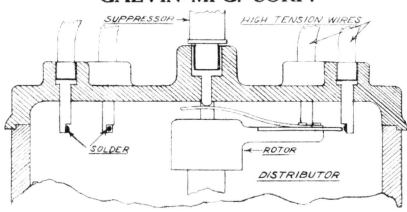

Figure 6

LINCOLN:

In the earlier model Lincoln that have the distributor coils mounted on the driver's side of the bulkhead, if it is impossible to eliminate antenna pick-up by any of the ordinary methods, it will be necessary to remove the coils and place them in the motor compartment. The same mounting holes may be used for the coils only they will be placed in the motor compartment instead of the driver's compartment.

The T junction of the flexible conduit should be loosened from the conduit and the flexible conduit placed up on the bakelite taper of the ignition coil. You will find enough slack in the flexible conduit to allow you to place the ignition cable proper in place before the flexible conduit is pushed up on the bakelite of the distributor. This will make a very neat appearing job and yet will accomplish the purpose desired.

On new Lincoln a dome light filter should be used - also it may be necessary to by-pass dome light feeder at the terminal board located back of the rear seat cushion with a .5 mfd. or larger capacity condenser.

LASALLE:

Remove the primary wire leading from the distributor to the ignition coil from the high tension conduit, keeping it outside this conduit. Shield the short length of wire leading from the distributor coil to the bulkhead, grounding this shield where it passes through the bulkhead. It will not be necessary to shield any wire other than this one.

In a few of the later custom models the application of two dome light filters will be necessary. They will have to be applied underneath the car at the junction boxes to their respective circuits.

On the 1932 model the coil is located on the bulkhead, on driver's side above the clutch pedals. To keep interference from being radiated by person driving car it is sometimes necessary to move coil to some other location.

OAKLAND:

For the reason of the No. 8 spark plug being located so close to the storage battery, the Oakland "8" presents a rather difficult installation problem. A shielding of the spark plug wire leading to the No. 8 spark plug will be of great assistance. It is extremely important that the "A" Battery connections of the radio be run directly to the post of the storage battery. The "A" Battery wires must be shielded clear up to the terminal posts, the shield covering the wire as close as it is practical to shield it.

It may often be necessary to place a double length of shielding over the "A" Battery wires as they come very close to the No. 8 spark plug.

Dome light filters need to be installed in all sedans and an additional generator condenser must be applied either at the starter connection to the bulkhead or at the ammeter to the instrument board.

Shielding must be placed over the lead from the distributor coil to the bulkhead, grounding this shield at the bulkhead.

PLYMOUTH:

The Plymouth, due to the motor floating in rubber, will need the motor bonded to the chassis frame in several places ... principally to the channel frame, again at the bulkhead, and again at the radiator. Braided shielding is recommended for this bond and enough slack should be left so that motor is free to float.

GALVIN MFG. CORP.

The spark control rod will need to be bonded where it passes through the bulkhead. The bonding of the control wires is ordinarily not necessary but may in a few special cases be necessary.

STUDEBAKER:

The 1931 and 1932 model Studebakers will only need one suppressor in the distributor, located in the distributor lead to the distributor. A heavy bond placed between the back cylinder bolt and the bulkhead of the engine will be necessary on all new models which are mounted in rubber.

Dome light filters should be installed in the dome light lead in all sedans.

Shielding of the distributor lead will not be necessary if the wire is in good shape.

CHAPTER II

SERVICE DATA ON MOTOROLA MODEL 7T47A:

Model 7T47A is a Superheterodyne designed exclusively for an automobile and consists of three type -39 tubes (which is the new screen grid Pentode Tube); two -37's and two LA tubes. The LA tube characteristics are identical to that of the -47 excepting it has 6.3 volt filament.

ELECTRICAL DESCRIPTION:

Figure #7 is a circuit diagram of this model. Following through the circuit diagram step-by-step and designating each tube's function should materially help the service man.

The antenna feeds the radio through a two contact plug whose purpose is to insure perfect connection of the antenna circuit. It is coupled to the grid circuit of the first radio frequency -39 tube, the feeder being balanced to the shielded antenna lead-in.

At the ground end of the grid circuit is a .05 condenser mounted directly on the coils. Its purpose is to complete the resonance circuit which is tuned by means of the first section of the variable condenser. This allows the ground end of the coil to be free for the introduction of negative potential from the Diode Tube.

In the plate of the first radio frequency tube is coupled the grid circuit of the detector oscillator tube, commonly known as the Autodyne Tube. This is an impedance coupled device. A small capacity, C-1, in conjunction with inductance L-6 is so matched to give a uniform gain throughout the broadcast band. This, of course, applies radio frequency into the control grid of the detector oscillator. The cathode of this tube passes through a small coil to the biasing resistors and its by-pass condenser. A by-pass condenser is used because it offers a high resistance to I.F. Frequency yet a comparatively low resistance to Radio Frequency. This, of course, keeps the plate circuit high impedance for I.F. and low impedance for R.F.

The plate circuit of this tube contains a standard I.F. Transformer with the exception that the primary resonance circuit has inserted in it a coil which couples with the cathode coil. This coil does not offer any appreciable impedance at I.F. Frequencies but constitutes a very close coupling for Radio Frequencies which thereby causes the tube to oscillate at radio frequencies. The frequency of oscillation being determined by the resonance circuit L-3-C-3, due, of course, to its close coupling to the cathode and plate coils.

These three coils are all wound and placed under a common shield, the I.F. Frequency then being imposed upon the grid of the I.F. Tube realizing full gain of the -39 tube when a resonance circuit is placed in the plate circuit of this tube. Energy in the plate circuit is transferred to the Diode Tube through capacity C-4. Where the Diode Tube acting as a simple half-wave rectifier builds up across the grid and cathode of the tube a negative potential depending upon the intensity of the I.F.A. tapped potentiometer is placed across the Diode Tube and highest negative potential is then fed back to the first R.F. Tube. A slightly lower negative potential is fed back to the grid circuit of the I.F. Tube, thereby giving automatic volume control action.

The fluctuation of the intensity of this rectified I.F. Frequency is tapped at the grid end of diode and I.F. Frequency is prevented from entering the audio circuit by means of a 250,000-ohm resistor, marked R-1. The regulation of audio is controlled by means of a 500,000-ohm volume control remotely located from the radio.

The plate of the first audio tube is impedance coupled through choke CH-1 to a push-pull input transformer, T-1. A .05 condenser is placed in the primary circuit to prevent DC from flowing through the primary winding. Choke CH-1, Transformer T-1 and .05 condenser are all contained in a common can.

GALVIN MFG. CORP.

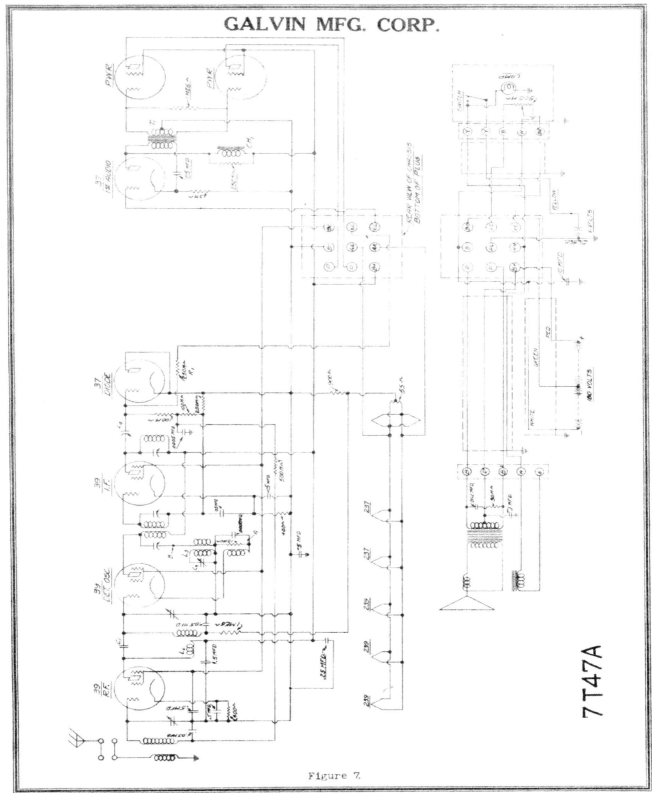

Figure 7.

7T47A

GALVIN MFG. CORP.

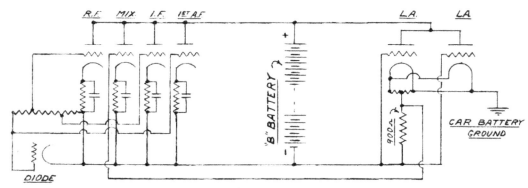

Figure 8

Figure #8 shows a theoretical lay-out of the output tubes and their methods of obtaining bias, also the Radio Frequency end and its method of coupling to the "B" Batteries, which illustrates the necessity of the variable condenser on the MOTOROLA being insulated. By this ingenuous method there is no chance of the output tubes being coupled with the Radio Frequency tubes.

TABLE Nº1						
	1ST R.F.	DET. OSC.	DIODE	I.F.	1ST AUDIO	LA's
E_p	176 V.	176 V.	0	176 V.	164 V.	176 V.
I_p	4.7 MILS.	13 MILS.	0	1 MIL.	4.1 MILS.	7.7 MILS.
E_g	2 V.	8 V.	0	2 V.	12 V.	18 V.
E_s	80 V.	80 V.	—	80 V.	—	176 V.
I_s	1.1 MILS.	.3 MILS.	—	1.2 MILS.	—	1.6 MILS. NO SIGNAL
NO SIGNAL I-CATHODE	5.8 MILS.	1.7 MILS.	.00001	5.5 MILS	4.1 MILS	
100-MMV. I CATHODE	.9 MILS	2.2 MILS	.0004	1.5 MILS.	2 MILS.	
E_f	6.2 V.	6.2 V.	6.2 V.	6.2 V.	6.2 V.	6.2 V
A-BATTERY VOLTAGE AT TERMINALS 6.25 VOLTS						

Volts cathode oscillating   8 V   } (Measured with 10 V 1000~/volt meter)
Volts cathode not oscillating 4 V   }

TABLE 1

Table #1 gives all the socket readings when using a 1000-ohm per volt voltmeter and should be applicable to any standard set analyzer. The readings, as you will notice, were taken (1) with the antenna connected and (2) when listening to a 100,000 microvolt signal. This should aid the service man in determining if he is getting proper automatic volume control action as indicated by the decrease in screen and plate current readings.

If set lacks sensitivity and if indications are towards oscillator not oscillating or not oscillating violently enough reference should be made to last part of Table 1, giving change in cathode volts when oscillating and not oscillating. The voltmeter must be placed at point A and B - as shown in Figure 7 and to stop oscillations point B must be grounded. To get at point A it will be necessary to loosen coil can located directly under oscillator section of variable condenser by removing two sheet metal screws that hold it to chassis, then tip the open end of coil can up high enough to insert test lead of voltmeter onto the coil lug connecting point A. Point B can be reached by inserting a screw driver at an angle in the right hand trimmer adjustment hole as this lug which is point "A", position one inch directly above the trimmer adjustment screw.

Figure #9 shows the power sensitivity curve of this model and the power sensitivity curve of the average home set which is put there so you may realize the difference in the shape of the power sensitivity curve of a radio that is not automatic volume control and the shape of the MOTOROLA.

GALVIN MFG. CORP

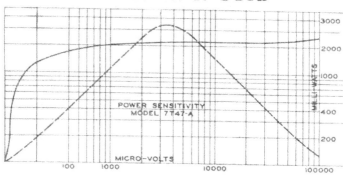

Figure 9

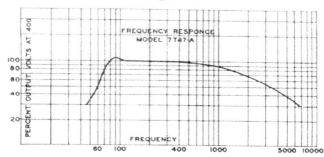

FIGURE 10

Figure #10 shows a frequency response of this model. You will notice the predominance at the low frequencies which accounts for the excellent tone received from this model.

The success of a Superheterodyne rests, to a large extent on the proper choice and use of an intermediate frequency unit. The frequency to which they are aligned, of course, is determined by the mechanical design of the variable condenser. The plates of the variable condenser used in our design are laid out mechanically to produce a frequency differential of 175 kilocycles and if the setting of the oscillator trimmer with re-spect to the radio frequency trimmer has been disturbed it will be necessary to realign. The realignment is accomplished by use of a 175-kilocycle oscillator. The circuit dia-gram of the oscillator is shown on the left hand side of Figure 11 with its proper application to a MOTOROLA.

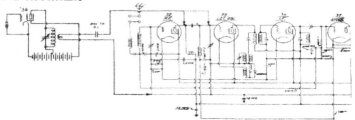

FIGURE 11

TESTING PRACTICE:

In case an oscillator set-up similar to Figure #11 is used, some means of reading the output of this oscillator must be provided. The oscillator may be modulated by insert-ing a bell ringing transformer at point "X" or the "B" supply may be taken from the 110 A.C. lines. Since most 175-kilocycle oscillators furnished with service kits are mod-ulated, the only problem is that of reading the output.

While the ear can be used for some purposes, it is not dependable and when there are so many output meters on the market it is folly to use the ear for any kind of service work excepting harmonic analysis. It is not recommended that intermediate frequency units be adjusted by the ear or by air test. The only safe and sure way is by means of the modulated 175-kilocycle oscillator fitted into either one of the two points shown in Figure #11, depending, of course, upon the strength of the local oscillator and by reading the output with an output meter across the plate terminals of the speaker.

GALVIN MFG. CORP.

Since the Motorola has no padder (left out purposely because of the stability of the cut-plate condenser) the whole secret of the oscillator being in track with the radio frequency depends entirely on getting a standard on which to start. Figure #12 shows a cut, giving the exact position in which to move the variable condenser so as to align the oscillator and second radio frequency trimmer to exactly 1400 kilocycles. If you align to this proper frequency and proper degree setting of condenser the oscillator will also track with the second radio frequency stage at any position of the condenser, assuming, of course, the previous paragraph dealing with intermediate frequency alignment has been accurately followed.

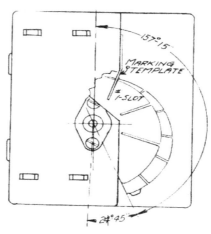

FIGURE 12

The antenna stage may need a little compensation but should balance itself perfectly on a dummy antenna representing 200 MMFD. of antenna capacity. If not, bending a condenser plate not over 1/32" will no doubt serve to compensate.

Every service man knows the value of isolating the trouble in a radio before starting to repair it. It should not be necessary to tell him that if the tone quality is bad, the first thing to do is check the output tubes and read their plate currents so as to get a suitable match. If that checks O.K. the following suggestion might be helpful.

Examine the speaker for rubbing voice coil, this being a quite common occurrence in all automobile installations as the speaker in auto-radios is exposed to a great deal more direct, dust, and mechanical vibrations than home set speakers and as a result speaker failures are a little more frequent in auto sets than in home sets. The examination for rubbing of voice coil requires a little practice and we suggest that you get the feel of the cone movement of a speaker known to be good and listen while moving to see if the voice coil is rubbing. Observe while testing this speaker known to be good how easily a voice coil can be made to rub by unequal pressure on the side of the cone. Therefore, while checking the speaker suspected to be bad, profit by the experience gained from the good speaker.

A rubbing voice coil sounds similar to two pieces of sand paper being very lightly rubbed together. If you are still in doubt the application of 50 volts 60 cycle across the two outside terminals of the output transformer, the two "B" terminals, will cause the speaker to pump sufficiently and if the voice coil is rubbing, noise will emit from the speaker instead of a perfectly free hum.

If the speaker sounds satisfactory see if the hum is equal on both halves of the output transformer and, if the speaker passes the above test, it is evidently not the cause of the trouble. A customary set analysis as to the bias readings, etc., should indicate the trouble.

All of the above tests can be simplified if the service man has a spare chassis, known to be good, or a spare speaker which can be substituted to quickly isolate the trouble.

Power supply, of course, is a very common cause of trouble and it is assumed that this factor had been checked into before going ahead with any of the above tests.

GALVIN MFG. CORP.

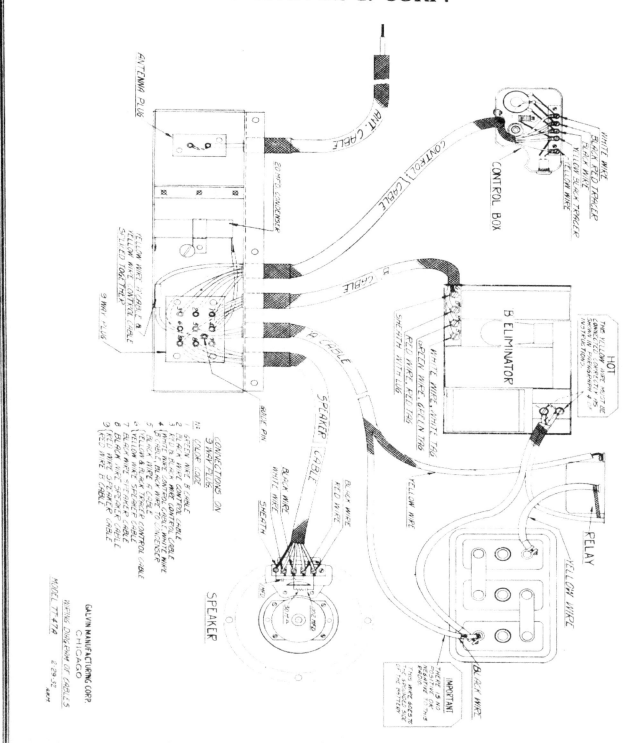

ANTENNA PLUG

ANT. CABLE

20 MFD. CONDENSER

WHITE WIRE
BLACK RED TRACER
BLACK WIRE
YELLOW BLACK TRACER
YELLOW WIRE

CONTROL CABLE

CONTROL BOX

YELLOW WIRE, B CABLE &
YELLOW WIRE CONTROL CABLE
SPLICED TOGETHER

9 WAY PLUG

B CABLE

THIS YELLOW WIRE MUST BE
CONNECTED CORRECTLY IS
SHOWN IN PARAGRAPH 4 OF
INSTRUCTIONS.

WHITE WIRE, WHITE TAG
GREEN WIRE, GREEN TAG
RED WIRE, RED TAG
SHEATH WITH LUG

HOT

B ELIMINATOR

A CABLE

BLACK PIN

CONNECTIONS ON
9 WAY PLUG

COLOR CODE
1 GREEN WIRE B CABLE
2 BLACK WIRE CONTROL CABLE
3 RED & BLACK WIRE CONTROL CABLE
4 WHITE WIRE CONTROL CABLE
5 WHITE, BLACK WIRE TO CONDENSER
6 BLACK WIRE A CABLE
7 YELLOW & BLACK TRACER CONTROL CABLE
8 BLACK WIRE SPEAKER CABLE
9 RED WIRE SPEAKER CABLE

SPEAKER CABLE

BLACK WIRE
WHITE WIRE
SHEATH

RED WIRE
BLACK WIRE

YELLOW WIRE

RELAY

YELLOW WIRE

BLACK WIRE

IMPORTANT
THIS WIRE GOES TO
THE GROUNDED SIDE
OF THE BATTERY.

THERE IS NO
POSITIVE OR THIS
NEGATIVE ON THIS
RADIO.

SPEAKER

30 M.A.
102 MFD.

GALVIN MANUFACTURING CORP.
CHICAGO

MODEL TT 47A   2-29-32 nm

WIRING DIAGRAM OF CABLES

GALVIN MFG. CORP.

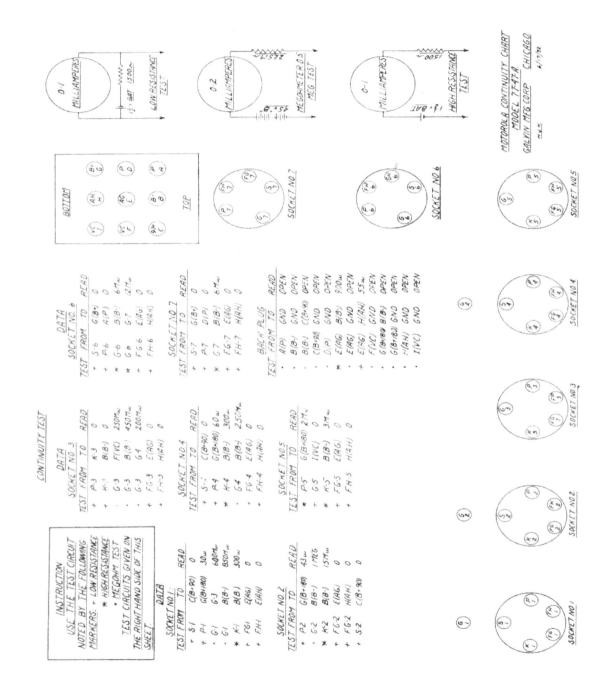

GALVIN MFG. CORP.

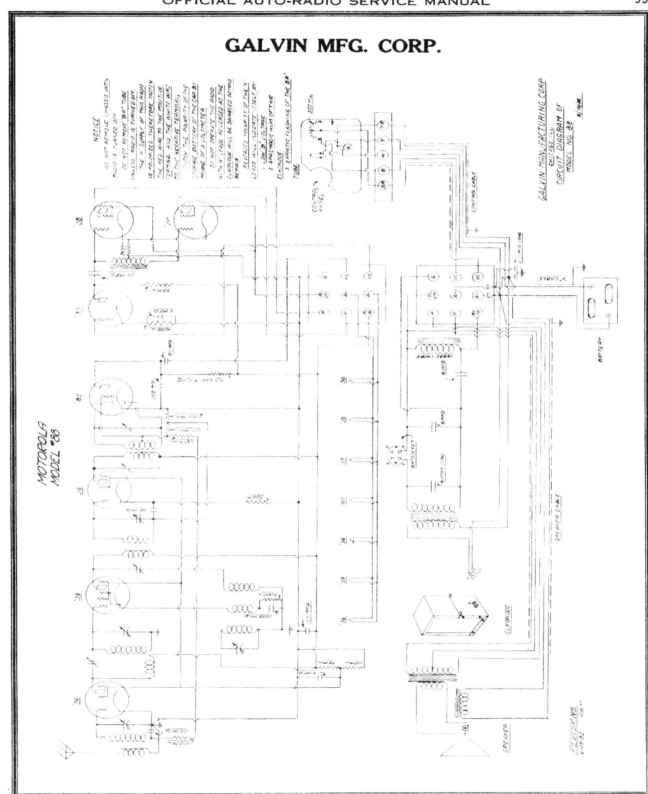

MOTOROLA MODEL #88

GALVIN MFG. CORP.

CHAPTER III

MOTOROLA ALL-ELECTRIC MODEL 88

The Motorola Model 88 is a 7-tube superheterodyne with a circuit similar to the Motorola Model 7T47A as described in Chapter II. The difference in circuit can be seen by referring to Figure (15), which gives a complete circuit diagram of the entire radio and power unit.

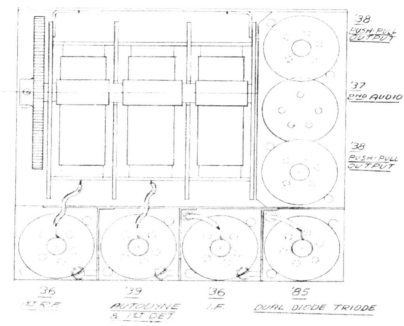

FIGURE 16

Figure (16) shows the tube layout and the sequence of tubes, reading from left to right as follows: 1st - 36 type used as 1st radio frequency, 2nd - 39 type used as an autodyne and 1st detector, 3rd - 36 type used as an I.F. stage, 4th - 85 type used as a Dual Diode Triode, meaning it is serving two purposes - that of the Dual Diode and a Triode, or three element 1st audio tube, 5th - 38 tube used as one of the output tubes operating as class "A" amplifier, 6th - 37 2nd audio tube, 38 - as the second of the push-pull output tube. The 36 and 39 tubes may be interchanged with each other, or all 36, or all 39 tubes can be used with the following expectations, when different type numbers are exchanged. It is recommended that a 36 be left in the 1st R.F. as you can notice by reference to Figure (15), where it does not have a *static bias, and if left disconnected from antenna over a very long period of time very short life can be expected of the 39 when used in that position and no increase in sensitivity will be noticed. Substitution of the 36 in the autodyne socket will result in a 5% decrease in sensitivity and a corresponding decrease in oscillator hiss. Substitution of the 39 in the I.F. stage is suggested when an increase in sensitivity is desired. It is perfectly safe to use a 39 in the I.F. stage as it is statically biased.

Ordinary AVC sets with normal tube hiss or noise level deliver sufficient bias to the tubes to prevent excessive plate current drain. The fourth tube, being second detector, is interchangeable with the 6 prong automotive type Wunderlich tube, Sylvania's 69 or 85 tube, or any other make of Dual Diode Triode. When the Wunderlich tube is interchanged with the 85 the grid clip which is normally connected on the top of the 85 tube can be ignored by merely taping it up and tucking it behind the tube, so that it will not become grounded. There is no substitution for the 5th tube, it being 37 second audio tube. The 6th and 7th tubes, or the 38 output tubes will be found to work best if the oval plate type of ER 38 is used, although any other type of 38 tube may be substituted.

* "Static bias is defined as self-biasing of the tube when there is no signal being imposed into the radio set, the radio being in a static condition."

GALVIN MFG. CORP.

Referring to the circuit diagram as shown in Figure (15) a description of the function of the radio frequency end of this set is described in Chapter II covering Motorola Model 7T47A. The changes in this model are fundamentally a Dual Diode second detector and the addition of one audio stage. The addition of the one audio stage is accomplished by the use of the Dual Diode Triode tube.

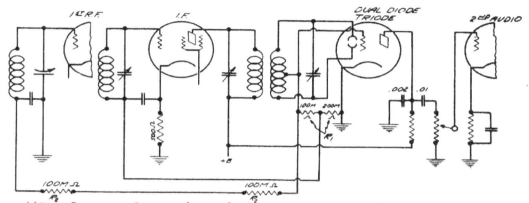

WIRING DIAGRAM SHOWING 'DIODE CIRCUIT' AND IT'S ASSOCIATED 'A.V.C.' LEADS

FIGURE 17

You will observe the difference in the diode circuit (Fig. 17) from that shown in the circuit diagram of the Motorola Model 7T47A as previously described in Chapter II, insofar that it is a full wave rectification of radio frequency. This is accomplished by the insertion of two small collectors that are located at the lower end of the cathode, and the result of the rectified radio frequency is built up across the 300,000-ohm biasing resistors marked (R-1) in the figure. The voltage produced across these resistors varies according to the modulation of the radio frequency and at this point it is fed into the grid of the triode section of the 85 tube; hence, it is amplified as shown in the diagram, Figure (15), as audio frequency. The resultant sum of this voltage built up across resistor (R-1) is filtered by means of the two resistors (R-2) into a steady D.C. source of negative potential which accomplishes the AVC action as described in Chapter II.

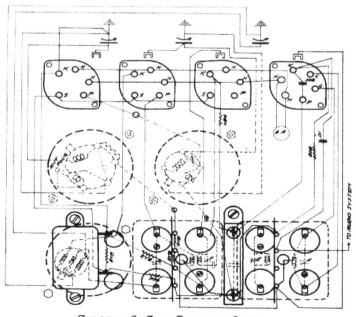

PICTORIAL OF RADIO FREQUENCY SYSTEM

FIGURE 18

GALVIN MFG. CORP.

Figure (18) - shows pictorially each resistor and condenser used in the set, as well as each terminal being shown phantomly its internal connection. This will describe to the service man thoroughly just what each terminal is in the front of model "88" and should assit him materially in continuity tests.

Figure (19) - shows the resistance audio coupling device that is used in the inverted socket which is a coupling medium that occurs between the 2nd audio and the push-pull stage.

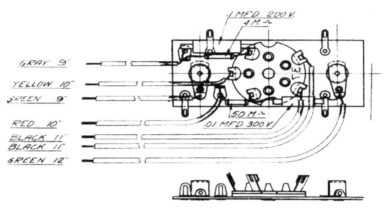

FIGURE 19

Figure (20) - is a continuity chart that is used for every component of the radio. These are made on standard continuity checks as shown alongside of the chart.

Table (2) - gives a socket analysis reading of the set with complete voltages and plate current of every tube, and these materially assist the service man in the elimination of defective units. Readings were all taken with a 1000-ohm per volt voltmeter, having a scale of 10 volts and 200 volts, 1 mill, 10 mills, or 100 mills.

TABLE Nº 2

	1ST. R.F.	MIXER	I.F.	DET.	2ND AUDIO	'38
I_p	2.5 MILS.	2 MILS.	2.5 MILS.	1.8 MILS.	.7 MILS.	8.5 MILS.
E_p	*180 V.	*180 V.	*180 V.	*30 V.	*38 V.	*200 V.
E_G	0	6 V	1 V.	0	2 V.	20 V.
E_S	*60 V.	*60 V	*60 V.			*200 V.
I_S	.7 MILS.	.3 MILS.	.3 MILS.			2 MILS
NO SIGNAL I CATHODE	3.4 MILS.	1.8 MILS.	2.8 MILS.			10.5 MILS
100 M. M.V. I CATHODE	0	2.3 MILS	.6 MILS			11.4 MILS.
E_F	5.8 V	5.8 V.	5.8 V.	5.8 V.	5.8 V.	5.8 V

TOTAL CURRENT OF SET 31.8 MILS. AT 185 V. 'B' MAX.
'A' BATTERY VOLTAGE - 6.5 VOLTS AT BATTERY TERMINALS
'A'   "   "   5.8   "   "   1ST R.F. TUBE.
* APPROX. VOLTAGES

TABLE 2

For the benefit of the service men that are equipped with a standard signal generator and after he has completed the set and is desirous of knowing if the automatic volume control is functioning properly, and if the set has a proper sensitivity, we are showing Figure (21) which gives the sensitivity of Motorola Model 88.

GALVIN MFG. CORP.

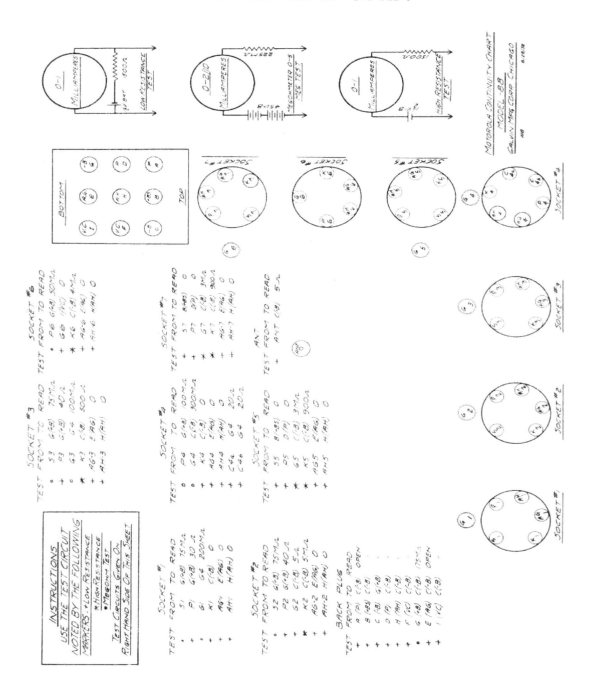

MOTOROLA CONTINUITY CHART
MODEL 88
GALVIN MFG CORP CHICAGO

GALVIN MFG. CORP.

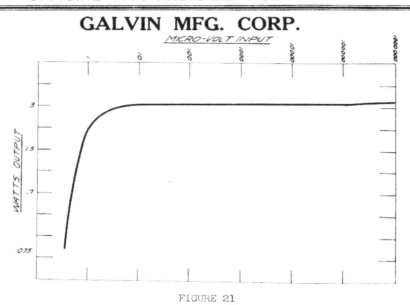

FIGURE 21

REMOVAL OF PARTS:

    The Model 88 receiver is simple and easy to service if it is properly dismantled.

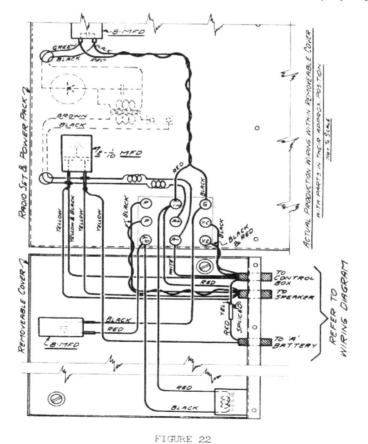

FIGURE 22

GALVIN MFG. CORP.

Inverted Tube Socket – – – If the inverted tube socket and its associated wiring becomes defective, and it is required to replace it, it is only necessary to remove the set from the housing and unsolder the two green wires from the dummy lugs located in the tube laying on the right hand side of the chassis – also remove the two volume control wires whose position under the terminal post is shown in Figure (15) the "B" plus wire, heater and ground, leading in the cable assembly, and when all of these wires are disconnected the entire cabling may be removed, or it may be replaced by a new one, or the old one repaired, which is wired as shown in Figure (19).

Removal of Diode Unit – – – If after thoroughly checking continuity of the diode unit it is found defective, it is only necessary to disconnect the four wires on each terminal, and the 5th wire coming out of the hole in the center of the unit. After the removal of these wires, the two nuts that hold the terminal strip should be removed, and the entire unit can be pulled out.

If the I.F. unit is found defective, the four wires should be removed from the terminals of the unit and the 5th wire coming out of the center should be removed from the by-pass condenser terminal. The two screws holding the terminal strip in place should be removed. The unit is then ready to be pulled out after the oscillator section has been removed as described below.

Removal of Oscillator Coil – – – The oscillator coil as shown in Figure (18) is located in the lower left hand corner of the chassis, and to remove it the tube shield should be removed by removing the three sheet metal screws holding the bottom of the tube shield in place and the two 6/32 nuts holding the back of the tube shield. It may then be lifted out of place, which allows the stator connection of the third variable condenser to be unsoldered. Also remove the black #20 wire that is soldered to the wiper of this same section. After removal of these two wires solder an additional 8" or 10" of wire on to each of these wires. This will act as a pull wire. Then remove the two hex-head screws located in the lower left hand face of the chassis which will release the coil and it may be removed and pulled out. After it has been removed, unsolder the two pull wires that were originally soldered on to the leads, removed from the variable condenser. These will be very important when you attempt to replace this unit, as it will put the wires back to the condenser in the same place they were removed. This pull wire will be very essential, because if the oscillator section is removed without placing this pull wire in place, you will find it necessary to remove all of the other coils in the radio in order to reassemble the oscillator grid and stator connections.

Removal of Antenna and Radio Frequency Coils – – – First remove the tube shield as previously described and unsolder all these stator connections on to the variable condenser. Remove the 160-tooth drive gear and remove the four hex-head sheet metal screws holding the variable condenser on to the brackets – then unhook the wipers from their position on the condenser and pull the condenser out, leaving the wipers soldered to the wires. This will allow complete access to the radio frequency and antenna coils.

Removal of Power Pack – – – Should the power pack become defective, it can be removed as a unit. It will be necessary to remove the housing from the car, or remove it from the bulkhead. Unscrew all of the screws holding the back cover plate in place tipping this back cover aside being careful not to pull any leads loose while working about it (see Fig.22). It will be found very convenient to use the middle mounting screw on the bottom which will align with the middle mounting screw of the back cover and by fastening those two points together the lid will be held in an out of the way position. All leads are amply long to allow it to rest in that position. Unsolder the brown and black #14 wires connecting the transformer to by-pass condenser, also unsolder the red (or green) and black wire leading to the 8 mfd. filter condenser. There will be no further wires necessary to unsolder. Remove the two screws holding the top of the transformer case located near these two red (or green) and black wires mentioned. Remove one screw holding the second side of the transformer case located on the right side of the outer housing, also the four screws, two holding the Elkonode and two holding the transformer located on the bottom of the outer housing. This will allow the Elkonode and the BR tube and transformer all to be pulled from the chassis as a unit. After it is removed, it can be tested by applying 6 volts to the large terminals with positive polarity to the brown wire and applying a 5000-ohm resistor across the red (or green) and black wires, an 8 mfd. electrolytic condenser and a voltmeter. With this setup the Elkonode unit should consume not more than 2.25 amperes and the voltage drop across the 5000-ohm load should be between 160 to 170 volts, provided the battery voltage is on exactly 6.3 volts. With the unit out in this manner and a bench setup made for testing any irregularities occurring in the power pack (and by reference to that section the circuit diagram shown in Figure 15) any trouble can be corrected.

It is not recommended by us that any repairs to the Elkonode be attempted by the service stations. All defective Elkonodes should be returned to the factory or the manufacturers of the Elkonode as indicated by the label on same.

Open Buffer Condenser – – – This condenser shown in Figure (15) as being applied directly across the secondary of the power transformer will be indicated by the failure of the BR tube to stay ionized. Ionization is the bluish-red glow always characteristic of Raytheon Rectifier tubes, while a shorted .05 condenser will be indicated by a spasmodic operation of the Elkonode, as well as failure of the BR tube to glow. As a general rule in all power packs when spasmodic operation of the Elkonode is observed, it is always an indication that the Elkonode is not feeding into the proper load. It is either unloaded or overloaded, and it is very hazardous if the Elkonode is allowed to operate in either one of the two conditions for any period of time.

GALVIN MFG. CORP.

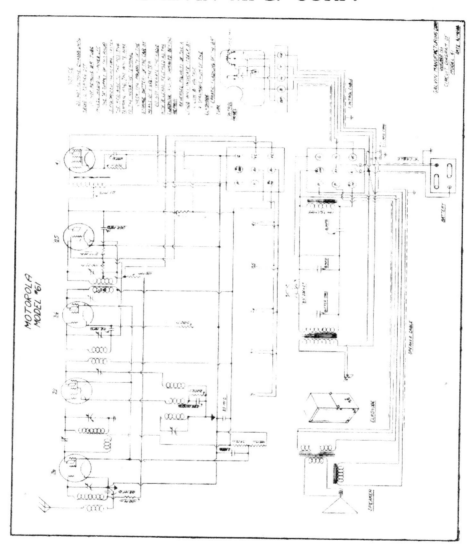

FIGURE 23

CHAPTER IV.

The Motorola Model 61 is very similar to the Model 88, the difference being in the design of the audio frequency end.

The Radio Frequency of Model 61 is interchangeable, serviced and wired in exactly the same manner as Model 88 as described in Chapter III. Therefore, all continuity charts, diagrams, and methods of parts removal can be taken from Chapter III.

Figure (23) shows the diagram of Model 61, and the difference in the audio frequency end can be readily seen by reference to Figure (15) of Chapter III.

The interchangeability of tubes as used in the radio frequency end can be done as described in Chapter III with all tubes, EXCEPTING THE 85 AND 41, which tubes must be interchanged only with tubes of corresponding numbers. The method of controlling volume in the Model 61 limits the adaptability of other types of detector tubes.

GALVIN MFG. CORP.

Fig. 1

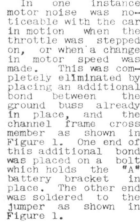

Fig. 2

FORD MODEL "V8" INSTALLATION

Because no suppressor can be effectively applied to the distributor lead, the Ford Model "V8" presents, more or less, of a perplexing problem for installation.

Motorola Engineers approached the problem from a different angle and after intensive research have simplified Ford car installations by the use of a Motorola "A" Filter (patent applied for).

Figure 1 shows the application of the "A" Filter at the "A" battery, which eliminates, in a big measure, the motor noise.

After the application of the "A" Filter there is a likelihood of antenna pick-up. This can be eliminated by removing the two 6-volt wires which are in the same conduit with the high tension wires. These two wires can easily be removed by cutting away the loom with a razor blade at its entry into the end of the conduit, then disconnecting the wire leading to the generator and the wire leading to the ignition coil, pulling them out and placing them on the outside of the conduit as shown in Figure 2. The primary ignition wire should be thoroughly shielded in the driver's compartment only.

This should effectively eliminate all motor noise in nearly every instance. In fact, it was so effective in many tests that spark plug suppressors were omitted, although it was not always reliable in all experiments. Note illustration in Figure 2.

In one instance motor noise was noticeable with the car in motion when the throttle was stepped on, or when a change in motor speed was made. This was completely eliminated by placing an additional bond between the ground buss already in place, and the channel frame cross member as shown in Figure 1. One end of this additional bond was placed on a bolt which holds the "A" battery bracket in place. The other end was soldered to the jumper as shown in Figure 1.

Installation men will find this practice is necessary in nearly all installations when the motor is mounted in rubber and is floating as free as it is in the Ford "V8" Model. When the battery ground bond is placed onto the motor block this allows a considerable potential difference to occur between the motor block and the channel frame. With the addition of another bond this trouble is neutralized.

Motorola "A" Filter can be had through Motorola distributors and dealers.

GALVIN MFG. CORP.

Fig. 3

MISCELLANEOUS NOTES

Here is a simple little trick for removing the radio chassis of Models 88 and 61 from the housing, after removal of two #10 screws holding bottom of chassis to housing.

See Figure 3. Grip the chassis with a pair of diagonal pliers at the point illustrated. A sharp jerk will easily remove the chassis from the housing as it puts a "pull" on the chassis in the correct position where it is held in place by means of a 9-way plug.

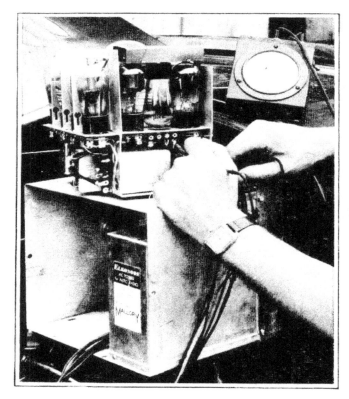

Fig. 4

Figure 4 illustrates the convenience of an extension cable when working on the removed chassis. If these extension cables are made up by the serviceman you will observe in the illustration Figure 4 the shielded volume control wires in this cable. This is very necessary to prevent audio squealing of the radio. By using the radio in the position shown, complete access to all parts may be had and is especially convenient for checking the plate current of each 36 tube. The drop from terminal B+200V on 9-way plug to plate terminal on socket should be approximately 3½ volts per tube.

GALVIN MFG. CORP.

MOTOROLA MODEL 6T12 PL
Galvin Mfg. Corp.
Circuit Diagram
August 15, 1931

GALVIN MFG. CORP.

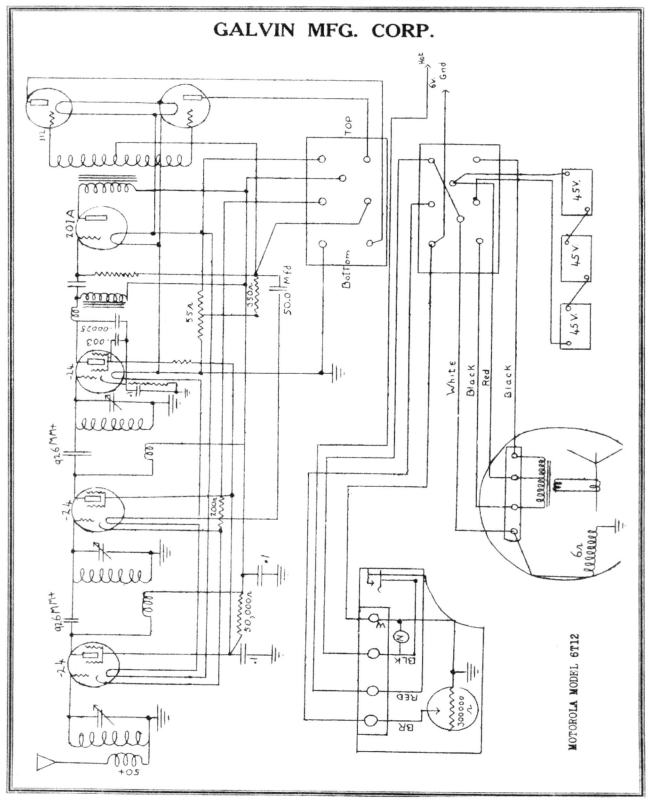

MOTOROLA MODEL 6T12

GALVIN MFG. CORP.

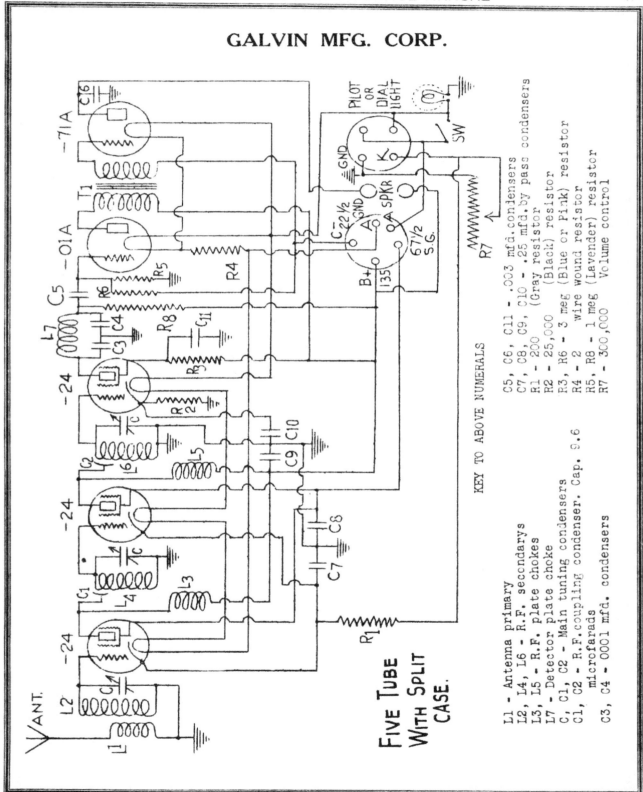

FIVE TUBE WITH SPLIT CASE.

KEY TO ABOVE NUMERALS

L1 – Antenna primary
L2, L4, L6 – R.F. secondarys
L3, L5 – R.F. plate chokes
L7 – Detector plate choke
C, C1, C2 – Main tuning condensers
C1, C2 – R.F.coupling condenser. Cap. 9.6 microfarads
C3, C4 – 0001 mfd. condensers

C5, C6, C11 – .003 mfd.condensers
C7, C8, C9, C10 – .25 mfd.by pass condensers
R1 – 200 (Gray resistor
R2 – 25,000 (Black) resistor
R3, R6 – 3 meg (Blue or Pink) resistor
R4 – 2 wire wound resistor
R5, R8 – 1 meg (Lavender) resistor
R7 – 300,000 Volume control

GALVIN MFG. CORP.

Motorola
7 Tube Set
Galvin Mfg. Co.
Schematic Diagram

Model 7738

GALVIN MFG. CORP.

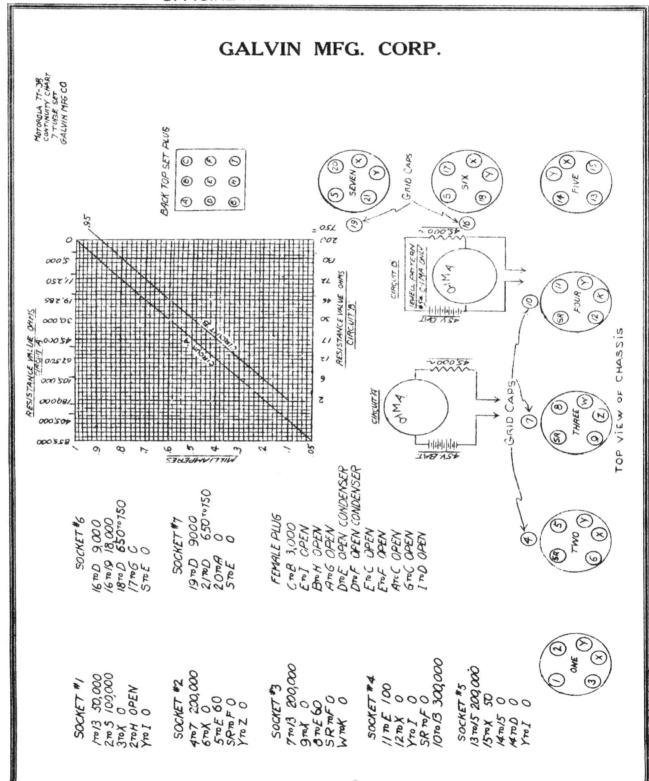

MOTOROLA TT-38
CONTINUITY CHART
7 TUBE SET
GALVIN MFG CO

BACK TOP SET PLUG

TOP VIEW OF CHASSIS

SOCKET #1
1 to 13 50,000
2 to 5 100,000
3 to X 0
2 to H OPEN
Y to I 0

SOCKET #2
4 to 7 200,000
6 to X 0
5 to E 60
SR to F 0
Y to Z 0

SOCKET #3
7 to 13 200,000
9 to X 0
8 to E 60
SR to F 0
W to K 0

SOCKET #4
11 to E 100
12 to X 0
Y to I 0
SR to F 0
10 to 13 300,000

SOCKET #5
13 to 15 200,000
15 to X 50
14 to 15 0
14 to D 0
Y to I 0

SOCKET #6
16 to D 9,000
16 to 19 18,000
18 to D 650 to 750
17 to G C
S to E 0

SOCKET #7
19 to D 9000
21 to D 650 to 750
20 to A 0
S to E 0

FEMALE PLUG
C to B 3,000
E to I OPEN
B to H OPEN
A to G OPEN
D to E OPEN CONDENSER
D to F OPEN CONDENSER
E to C OPEN
E to F OPEN
A to C OPEN
6 to C OPEN
I to D OPEN

GALVIN MFG. CORP.

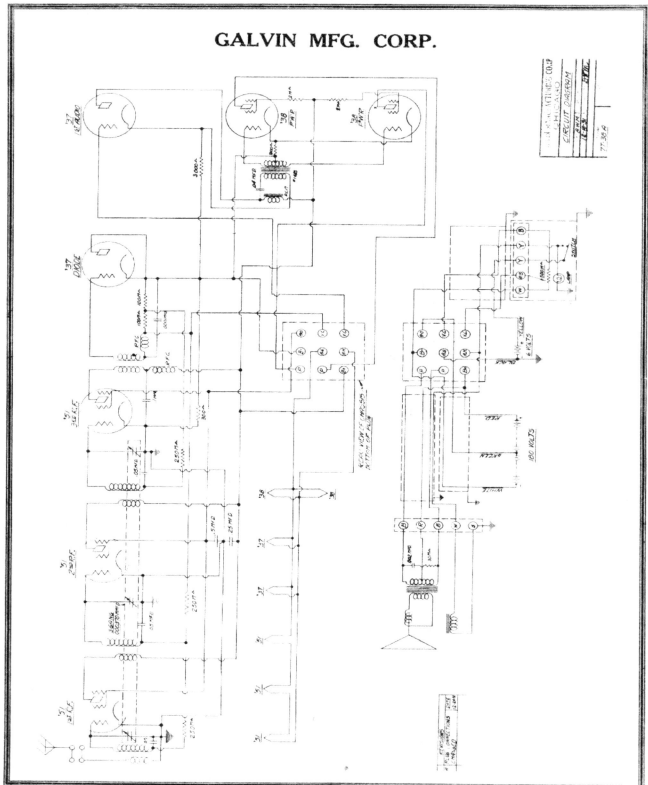

GALVIN MFG. CORP.

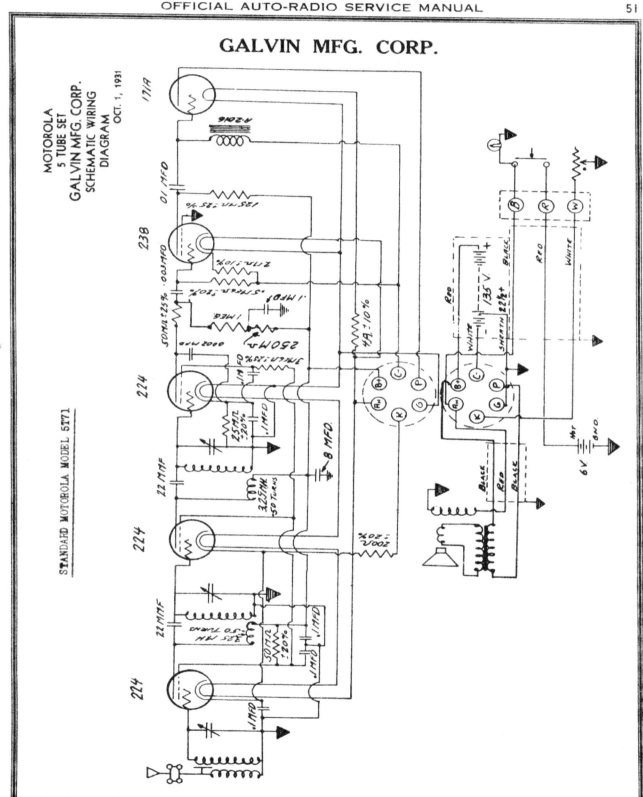

MOTOROLA
5 TUBE SET
GALVIN MFG. CORP.
SCHEMATIC WIRING
DIAGRAM
OCT. 1, 1931

STANDARD MOTOROLA MODEL 5T71

GALVIN MFG. CORP.

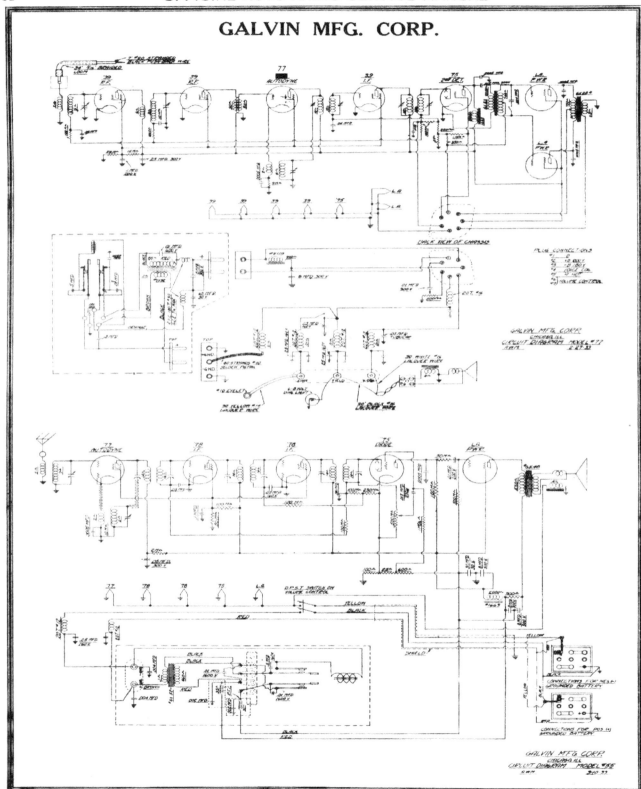

GALVIN MFG. CORP.

Motorola Auto Radio INSTALLATION and SERVICE NOTES

Supplement to Motorola 1932 Service Manual

Series A	GALVIN MANUFACTURING CORPORATION - CHICAGO, ILLINOIS	No. 353

ta on previous Motorola Models will be found in Motorola 1932 Service Manual.

Information on Installation and Elimination of Interference contained in the 1932 Manual is applicable to 1933 Motorola Models.

General Installation Instructions are packed with each set. Tags attached to the cables of both models, and labels affixed to the "B" Power Supply contain IMPORTANT INSTRUCTIONS. Be sure you read these instructions carefully before installation is attempted.

Supplements will be issued from time to time. So they may be available as a unit, it is suggested they be filed in a loose leaf binder.

MOTOROLA MODEL "77"

Motorola Model "77" is a seven tube superheterodyne which may be mounted on the bulkhead either in the motor or driver's compartment. A list of set locations for most of the 1932 models is given on page five of the 1932 Motorola Manual. Locations in 1933 models are similar. Model "77" is mounted either in an upright or inverted position, BUT SHOULD NOT BE MOUNTED SIDEWAYS as serious damage can result. (NOTE THE INSTRUCTIONS ON TAGS AND STICKERS PACKED IN EACH SET).

ANTENNA   An antenna of at least nine square feet is recommended for all Motorola Models where roof type antenna is used. Information on all types of antennas is given on pages 6 and 7 of the 1932 Motorola Service Manual.

NOTE...All cars equipped with antennas should be carefully checked for possible grounds by connecting an 0 to 50 or 0 to 100 volt high resistance volt meter in series with a 45 volt "B" battery and connecting these from the antenna wire to the metal frame of the car. If any deflection in the meter occurs, the antenna is grounded and the ground must be cleared before using the antenna.

SERVICING   The chassis and eliminator on the Model "77"
MODEL "77"   are plug-in type which may be removed as a unit. All parts at the rear of the set housing may be reached by removing the rear cover plate.

Circuit diagrams show relation of the connected plugs so the chassis eliminator and outside housing assembly may be checked individually if desired.

ALIGNMENT OF THE INTERMEDIATE FREQUENCY TRANSFORMERS

The I.F. transformers in the Model "77" are tuned to 456 kilocycles, a standard test oscillator calibrated to 456 kilocycles and an output meter is required for alignment. The output meter should be a 0 to 10 volt 1000 ohm per volt voltmeter, or the Motorola Utility Meter connected

from Terminal No. 6 on the chassis plug to ground. The output of the oscillator may be fed into the grid of the autodyne tube and I.F. trimmers adjusted for maximum reading of the output meter.

ALIGNMENT OF THE VARIABLE CONDENSERS

The alignment of cut plate variable condensers, the type used in Model "77" differs from the alignment of the variable condenser with a padder, in that the cut plate condenser has a fixed mechanical ratio between the capacities of its sections. In the past it has been possible with padders to align the condenser with an accuracy of ten degrees of rotation of the condenser plates - that is, it could be set at the high frequency end with all trimmers in alignment and then could be re-aligned at the low frequency end by rocking the condenser while adjusting the padder, thereby finding the point of proper alignment. This procedure cannot be used with a cut plate condenser.

The simplest and easiest way to align a cut plate condenser is as follows: Use a standard test service oscillator and output meter. Connect a 200 mfd. condenser in series with the antenna lead of the oscillator and connect to the antenna of the radio set. CAUTION: Before proceeding be sure that the I.F. transformers have been tuned to exactly 456 kilocycles. See paragraph 5. This is absolutely necessary otherwise the proper alignment of the variable condensers can never be attained. After assurance that the I.F.'s are in correct alignment, set the test oscillator to approximately 1400 kilocycles and apply this energy to the antenna post of the radio set. If this frequency is accurately known you can get approximately the correct starting position by setting the pointer on Model "77" to the indicated frequency, however if it is not known this is not essential.

Align all three trimmers to 1400 kilocycles. Then move the variable condenser to approximately the 600 kilocycle position and check the alignment of the second radio frequency trimmer. If it is found that the trimmer must be moved either in or out to return to resonance it is an indication that the variable condenser is not at correct starting position for the initial setting of the test oscillator. If, for example, it is found that the trimmer must be screwed down, it is an indication that the radio frequency tuning condenser requires more capacity at the low frequency end. Therefore, return to the initial high frequency setting of the condenser. Change your test oscillator to correspond with this setting of the condenser. It is not necessary to return to the exact setting you originally had. Re-adjust the second radio frequency trimmer which was moved when it was in the low frequency position. This will restore it to its initial setting of the oscillator trimmer.

Remember, that the second radio frequency condenser needs more capacity at the low frequency end so it is necessary to move the condenser a few degrees inward, which gives more capacity to this condenser, leaving the test oscillator in the same position. Screw the oscillator trimmer until the signal is brought back, then go over all three trimmers to assure yourself that they are in perfect alignment. Move the variable condensers back to approximately 600 kilocycles and re-check the second radio frequency trimmer the second time, and if the condenser had been moved sufficiently while you were at the high frequency end the R.F. trimmer will show resonance. If it was moved too far it will be indicated by having to move the radio frequency trimmer out instead of having to tighten it, as was necessary in the first trial.

After having found the proper starting point so that the second R.F. and oscillator trimmers are in alignment, the antenna stage should fall in exact alignment with the second radio frequency condenser. If it does not it may be necessary to bend the end plate sections slightly in order to align it with the second R.F. tuning condenser.

GALVIN MFG. CORP.

In the above set-up caution should be taken to see that the points chosen in which to align the radio set are in channels that are not occupied by a local broadcast station. This often upsets the measurements and you find you are tuning to the heterodyne beat occurring between your local test oscillator and the local broadcast station. This, of course, will tend to give a double peak.

"B" POWER SUPPLY

Model "77" uses a self-rectifying Elkonode which eliminates the rectifier tube used in former Motorola all-electric models. The yellow "A" lead of the "77" may be connected to any point on the electrical system of the car....ammeter, starter button or battery.

It is necessary to maintain a definite polarity at the Elkonode. For this purpose a polarity changing switch has been provided at the rear of the set housing. The polarity is indicated through a small hold at the lower right rear corner of the set housing. If a red disc appears in the window which reads plus (+) ground, it means that the "B" supply unit is set to be used in cars having the positive side of the battery grounded. If a black disc appears which reads minus (-) ground, it means that the "B" supply unit is set to be used in cars having the negative side of battery grounded. Be sure to determine exactly which side of the car battery is grounded. Then be sure that the marking on the indicator corresponds with it. To change the polarity proceed as follows:

(1) Remove "B" supply unit by prying with screw driver in the slots provided on either side of the "B" power unit.

(2) After removal of the "B" power unit you will observe two receptacles on the rear partition - one on the left and one on the right. The one on the left side requires no adjustments but the one on the right side may be moved up or down in its slot.

(3) Insert a small shank screw driver or ice pick in one of the jacks of this receptacle and adjust up or down for desired indication in window.

(4) Replace "B" power supply.

MAKES OF CARS HAVING "POSITIVE" GROUND - Marmon - De Soto - Cadillac - Pierce-Arrow - Dodge - Packard - Graham - Plymouth - Studebaker - Auburn - Hupp. - Franklin - Rockne - Ford - Chrysler - Nash Twin Ign.

MAKES OF CARS HAVING "NEGATIVE" GROUND - Reo - Chevrolet - Stutz - Willys-Overland - James Cunningham - Lincoln - Continental - Buick - Oldsmobile - Pontiac - Hudson - Essex - Nash Single Ign.

For any cars not listed phone nearest car distributor or dealer.

Access may be gained to the interior of power supply for service by removing the round head screws which hold the bottom cover plate and remove this plate.

It will be noted that the connections to the Elkonode are made by means of a floating socket and to replace, it is only necessary to pull the Elkonode out of the socket.

CAUTION: When replacing Elkonode make sure that it lies with the label either down or up, but not on the sides. This is extremely important for if placed on the side the vibrating reeds will pull against gravity and their life will be shortened.

REMOVAL OF PARTS FOR REPLACEMENT

Almost all the parts of the chassis assembly, "B" power assembly and outer housing assembly may be removed for replacement without disturbing any other units. There are several, however, which cannot be removed individually. Therefore, to remove the antenna coil, the second R. F. coil or the oscillator coil, it will be necessary to remove the tubes from the chassis and remove the tube shield which is held in place by two sheet metal screws. The screws holding the coil cans may now be reached and removed.

To remove the I.F. transformer it will be necessary to remove the transformer mounting bracket to which the I.F. unit is attached.

To remove the diode feeder, loosen the transformer mounting bracket and it can be moved sufficiently to get at the screws holding the diode feeder unit. To remove volume control unit (located in the rear of the set housing) remove screws holding worm gear bracket and volume control bracket. Disconnect all leads to the switch and volume control. The volume control assembly may now be removed and replaced with a new unit, care being taken in re-assembling.

All bypass condensers except the R. F. plate bypass are of the tubular type and are set in thimbles in the chassis. Should any one prove defective it may be pushed out and replaced.

MOTOROLA MODEL "55"

Motorola Model "55" is a 5 tube superheterodyne. The chassis, "B" power supply, and dynamic speaker are assembled in one unit. The "55" is so designed that all component parts are assembled on the speaker plate. By removing all screws except the six hexagon head screws and four round head screws located at the edge of the speaker drill, the entire set may be dropped out of the outer housing for servicing or tube replacement.

SERVICING MODEL "55"

After removing the outer housing, the chassis can be inverted and the six hexagon head screws removed. After removing these screws, the speaker plate and speaker may be lifted off and placed at the side of the chassis without disconnecting the speaker wires. After this has been done, all wiring will be exposed and easily accessible for service.

Care should be used in re-assembling so the speaker wires do not get pinched under the speaker "pot".

Reference to the circuit diagrams, will show that resistance values from the various parts are given so the set may be completely analysed by resistance method.

ALIGNMENT OF I.F. TRANSFORMERS AND TUNING CONDENSERS

The method of aligning the I. F. transformers at 456 kilocycles and the alignment of the gang tuning condensers is the same as used in Model "77". See Page ( ), Paragraph ( ).

In the alignment of I. F. transformers it will be noted that the third I. F. transformer or diode feeder may be reached with a non-metallic screw driver inserted in the hole provided in the upper part of the chassis located between the first I. F. transformer and the "B" power supply housing. This screw driver may be a piece of 3/16 dowel rod 10 inches long to which is fastened a small metal screw driver tip.

Any of the various units of the Model "55" may be removed individually for repairing or replacement without disturbing other units.

"B" POWER SUPPLY

The self-rectifying Elkonode is also used in the Model "55". This along with the power transformer are housed in a single unit in the upper right corner of the chassis. The complete power unit may be removed by disconnecting the two "A" supply leads at the power unit terminal strip and disconnecting the B- (minus) and B+ (plus) leads at their respective terminals located on the set chassis. Should the Elkonode require replacement it may be removed by disconnecting the four wires extending from its sponge rubber housing and connecting the new unit as shown in Figure ( ).

CAUTION: Do not attempt to make any adjustments to Elkonode.

GALVIN MFG. CORP.

Motorola Auto Radio INSTALLATION and SERVICE NOTES

Supplement to Motorola 1932 Service Manual

Series A	GALVIN MANUFACTURING CORPORATION—CHICAGO, ILLINOIS	No. 243

INTERFERENCE

A general discussion of ignition interference will be found on Pages 3, 8, 9, 10, 11 and 12 of the Motorola 1932 Service Manual.

There are two types of interference. These are commonly called Chassis and Antenna pick-up.

Briefly, chassis pick-up is motor interference heard through the loud speaker with the antenna of radio disconnected. It is an indication that waves set up by the motor and ignition system are passing through the radio and being amplified as radio frequency. Due to its design and construction chassis pick-up is rarely encountered in Motorola.

Often chassis pick-up motor noise comes into the set through the "A" lead. A Motorola Dome Lite Filter used at the point of the "A" lead attached to the "A" battery of the car whether at the battery or the starter usually overcomes this noise entirely.

If chassis pick-up still occurs in the model "77" it is due to either of the following causes: (1) Defective condenser in the Eliminode system. Check by replacement with another condenser of the same value. (2) Cover of set making poor ground to set housing. Remove cover and clean lips of cover and set housing with fine sand paper.

If chassis pick-up occurs in Model "55": (1) Use Motorola Dome Lite Filter. Connect one side to the battery circuit of the car and the other to the end of the yellow wire. Connect battery condenser wire to the ground. Motorola Dome Lite Filter is a standard item which can be supplied by the distributor at 60¢ each list. (2) If high tension coil is located on instrument panel shield high tension wire from coil to bulkhead, grounding this shield at the bulkhead. In some cases it may be necessary to cover the head of the ignition coil with a metal shield.

ANTENNA PICK-UP

Interference reaches the antenna through the following methods: (1st) Direct Radiation. (2nd) Conduction and Radiation. (3rd) Eddy Currents in surrounding metal.

DIRECT RADIATION

This type of interference is radiated directly from the ignition system, high tension wires, coil distributor, etc. Spark Plug Suppressors reduce the radiation from the spark plugs and high tension wires. Spark Plug Suppressors do not, however, eliminate the radiation from the coil and distributor proper. As the distributor is usually well shielded by the hood of the car it is not necessary that any steps be taken to shield it further, other than to make sure that the hood is making good contact to the hasps that hold it in place.

In cars such as the 1932 "V8" Pontiac, where the distributor is mounted close to the wooden floor board it is sometimes necessary to tack copper screening to the floor boards and ground it to the frame of the car. This prevents the distributor from radiating directly either to the antenna or to be picked up by the feet of the driver or passenger of the car and to re-radiate it to the antenna.

The coil when mounted under the hood may be considered as being as well shielded as the distributor. Also, as in the case of the distributor, where the coil or high tension wire leading from it to the distributor comes close to the wooden floor boards a copper screening should be tacked on the floor board to prevent the occupants of the front seat from picking up the interference when they have their feet in a position near the coil. Chevrolet is a good example. Cars with coils mounted on the dash or on the bulkhead in the driver's compartment are likely to radiate to the antenna.

Where the coil is mounted on the bulkhead in the driver's compartment the most simple and positive remedy is to remove it in the motor compartment. It can usually be mounted in the motor compartment directly opposite the position it assumed in the driver's compartment and the same mounting holes used. Examples of cars on which this is necessary, are Lincoln "8" and LaSalle 1932 models.

Coils mounted on the dash are difficult to move to the motor compartment due to the fact that the ignition switch is mounted in the base of the coil. For this reason a separate ignition switch would need to be used if the coil were removed. Usually in the case of dash mounted coils it is sufficient to shield a high tension lead that runs from the coil to the distributor. The most satisfactory way to do this is to slip a piece of loom over this high tension wire, and then slip braided shielding over the loom, keeping the ends of the shielding one or more inches from the coil and distributor. The loom increases the gap between the ignition wire and shielding and not only makes the shielding more effective, but prevents leakage that might affect the ignition system.

This shielding should be grounded to the bulkhead at a point where it enters the driver's compartment. Some coils, however, are very violent radiators and it is necessary to go further and partially shield the coils themselves. This may be done best by placing a Metal Shield Can (see Figure A) around the front of the coil. Drill hole in can to admit the high tension lead and slot the sides of the can to admit the low tension wires. Flare the end of the can, turning the ends of the flares up. Wrap wire around the tips of these flares and draw it tight so as to hold the shield securely in place.

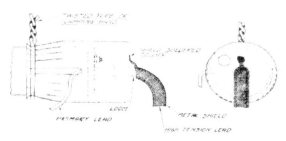

(Fig. A)

CONDUCTION AND RADIATION

Interference reaches the antenna through being induced in wires that come close to high tension radiators. These wires may then conduct the interference to a point near the antenna. A striking example of this is the dome light lead. The dome light lead in many instances runs very close to the antenna or to the antenna lead. So common is this type of interference that it should be one of the first things checked when interference is encountered. The best method of checking it is to cut this lead at a point four or five inches from where it enters the front corner post. It should then be left disconnected until all other types of interference have been eliminated. Afterwards it may be reconnected. If it causes interference when reconnected a Motorola Dome Lite Filter should be connected in series at the point where the lead was clipped. In some cases the ground lead of the dome light will be carrying interference up to the antenna. This may be eliminated by transferring the point where this lead is grounded to a point on the frame that is not carrying ignition currents. Various wires that run in the motor compartment or near the coil will pick up interference and carry it to some point near the antenna and re-radiate to the antenna or lead-in. Connecting a .25 to a 1 mfd. condenser from the ammeter or from the individual wire to ground will usually eliminate this type of interference.

GALVIN MFG. CORP.

In cases of an under-car antenna the tail light and spot light leads are often a source of interference. By passing these with a .25 to a 1 mfd. condenser is often sufficient, but the Motorola Dome Lite Filter inserted in series with the wire is usually a positive cure.

EDDY CURRENTS IN SURROUNDING METAL

As the body and frame of the car act as a return path to the battery for ignition currents and as this metal is of a type which offers considerable R. F. resistance, it is subject to very strong eddy currents. The metal corner post up which the lead of the antenna runs and the metal frame around the top of the car are the most troublesome objects for eddy currents. The simplest and in most cases the complete remedy for this is to shield the antenna lead from a point four to six inches from the antenna proper and down to the shielded loom of the antenna lead to the set so as to make a continuous shield from the set to the top of the corner post of the car. The shield should be grounded at this point near the antenna to the metal frame work at the top of the car. Use a wire insulated as heavily as possible so that the shield is kept far away from the antenna lead. The capacity between the two sometimes causes eddy currents to be induced in the antenna lead shield itself by its coming in contact with the dash, which frequently carries violent eddy currents that are induced in the wire.

For this reason it is very good practice to check whether the interference is less or greater if this shield grounds to the dash. If grounding to the dash increased the noise it will be necessary to insulate the shield with friction tape where it comes in contact with the dash. Grounding the antenna lead shield to a point on the bulkhead is often helpful in reducing eddy currents. The 1931 and 1932 Chevrolets are excellent examples of where it is often necessary to ground the shielded lead in to the metal frame at top of the car and insulate it from the dash.

Further steps may often be taken to reduce the flow of these eddy currents through parts of the car body that come near the antenna. The best remedy for this is to supply a shorter and lower resistance path through which ignition current may return to the storage battery. Bonding the motor to the bulkhead and frame of the car with heavy flexible bonds will usually accomplish this. In cases of a floating power type motor and where the motor is mounted on rubber this bonding is often absolutely necessary. In cars such as the Chevrolet models where the car battery is grounded to the transmission or motor it is sometimes necessary to run a flexible bond to the channel frame of the car from the grounded side of the battery. Braided copper shielding not less than 3/8" wide is the best type of bonding for this purpose.

The metal loom of various controls that lead from the dash into the motor compartment sometimes must be bonded to reduce eddy currents and radiation. The spark control loom, throttle loom, etc., may be bonded to the bulkhead at points where they enter the driver's compartment. In soldering the loom of these controls be careful not to solder the inter control wire to the loom. The best way to prevent this is to have some one operating the control continuously while the loom is being soldered.

ACCUMULATIVE DISCHARGE

This type of interference is one of the most peculiar types encountered. It appears as a spasmodic discharge and is very similar to atmospheric static in its irregularity.

The cure is a .002 to .006 mica condenser connected directly across the primary breaker points of the distributor. This condenser gives power factor control to the paper condenser, which is already across the distributor breaker points as part of the ignition system. This mica condenser tends to make the paper condenser much more effective in reducing interference. It cannot affect the ignition system in any way.

Before concluding it might be well to point out that the success in the elimination of ignition interference is governed by common sense procedure. First, carefully examine the car on which the radio is installed or is to be installed and try to decide which are the most likely causes of interference in that particular type of car. Don't overlook a possibility for noise. Then proceed in a step-by-step manner in the order of their importance. Always bear in mind that the interference may be coming from several different sources and that unless all of these are eliminated at the same time you will not have accomplished anything. Elimination of one of the sources of interference and leaving the others will as a rule have no apparent effect.

The quickest method of eliminating possible source of interference is to equip yourself with a number of lengths of flexible test bonds which are equipped with storage battery clips at either end. Also equip yourself with a few good grade non-inductive 1 mfd. by-pass condensers. Equip the leads of these condensers with test clips. This equipment makes it possible to quickly clip on test bonds or condensers at points that are possible sources of interference. Then as each possible source is eliminated by this method be sure to leave the test bond or condenser in place until interference has been entirely eliminated. After the interference has been eliminated, the various test bonds and condensers may be removed one by one. When the removal of a bond or condenser causes the interference to reappear, a permanent one should be mounted in its place.

* * * * *

As a handy list the following suggestions are given for the suppression of ignition interference. In this list we have endeavored to give these hints in the order of their importance:

(1) Apply suppressors to spark plugs and distributor.
(2) Apply generator condenser.
(3) Re-route primary wire from coil to distributor, keeping it as far as possible away from high tension wire.
(4) Connect Motorola Dome Lite Filter to dome light wire at point where it enters front corner post.
(5) Shield high tension wire if coil is mounted on instrument panel.
(6) Shield antenna lead-in wire from radio set to top of front corner post. Ground shield at both ends.
(7) Shield primary wire from coil to distributor.
(8) Connect a .002 to .006 high grade mica condenser directly across the primary breaker points of the distributor.
(9) Bond the upper metal parts of the car body to one another and return a heavy copper bond from these points down to the bulkhead of the car. (This is usually necessary in cars using composite wood and metal body construction).
(10) Bond where necessary all control rods and pipes passing through the bulkhead.
(11) Shield head of coil when mounted on instrument panel as shown in Fig. A.
(12) Cover floor boards of car with copper screening.
(13) Adjust spark plug points to approximately .028 of an inch.
(14) Clean and adjust primary distributor breaker points.
(15) In cars having rubber motor mountings connect heavy bond from grounded side of battery directly to frame of car.
(16) Connect a .50 to 1 mfd. condenser from hot primary side of ignition coil to ground.
(17) If ignition coil is mounted on driver's side of bulkhead move it to the motor compartment side using the same holes for mounting.
(18) Clean ignition system wiring. Clean and brighten all connections. Replace any high tension wiring having imperfect insulation.
(19) Ground metal sun visor and rain troughs if necessary.
(20) Make sure hood of car is well grounded. Clean hold-down hasps on both sides.
(21) Ground instrument panel and steering column to bulkhead.
(22) When under-car aerial is used connect a .50 mfd. condenser to tail and spot light wires.

GRIGSBY-GRUNOW CO.
TECHNICAL DATA PERTAINING TO MODEL 66 AUTO RADIO

CIRCUIT

The circuit is largely conventional but has some unique features. It is a superheterodyne using an intermediate frequency of 175 K.C. It is very selective and free from images and tweets, since it uses a 3-gang tuning condenser. The circuit sequence is as follows: One stage of tuned radio frequency amplification, composite modulator and oscillator, one stage intermediate frequency amplification, diode detector, one stage of low level audio amplification followed by the power output stage. Full automatic volume control on three tubes is obtained resulting in an excellent characteristic with respect to input signal voltage.

DELAYED AUTOMATIC VOLUME CONTROL

The audio detection and Automatic Volume Control are obtained from the diode circuit by the "delayed rectification method". In this method the diode plates operate at somewhat negative bias so that no A.V.C. results until a certain signal level has been reached. This results in much higher outputs at low signal levels than in the ordinary methods of A.V.C. since the set is left in its most sensitive condition until reasonable power output has been obtained. In the old methods of A.V.C. any input signal at all starts to decrease the sensitivity of the set.

TUBES

The tubes used in the Model 66 represent the latest advance in the art. The G-6E7 used as a radio frequency and an I.F. amplifier is a screen grid tube of characteristics somewhat similar to the type G-58 used in home receivers. It is spray shielded and the spray is connected by itself to one of the prongs of the base. This allows the shield to be directly connected to ground rather than to the cathode as heretofore. This is a definite advantage since when the spray is connected to the cathode it must be carefully insulated from the chassis pan in most cases, due to the fact that the cathode is not operated at ground potential.

The G-6A7S used in the composite detector-oscillator position is the new Pentagrid converter recently developed. This tube presents definite advantages over tubes previously used in this service, in that the automatic volume control may be allowed to operate on it, thereby allowing a far better degree of control than heretofore.

The G-6C7 is a double diode triode similar to the G-55 and G-75 types but having an amplification factor intermediate between them. It has the advantage over the G-55 of having considerably more gain and has none of the power handling deficiencies of the G-75.

The G-6Y5 rectifier tube is a full wave, spray shielded, mercury vapor rectifier. Its use materially cuts down the amount of high frequency interference between the "B" supply and the receiver.

"B" SUPPLY

The power supply system used in this receiver is unique in several respects. The MAJESTIC Duro-Mute vibrator as used in the Model 116 series has been re-developed and enlarged to handle adequately the power demands of the receiver. The vibrator armature, though smaller in overall physical size, has a far greater electrical capacity than heretofore, the diameter of the circuit-breaking contact points being materially increased.

GRIGSBY-GRUNOW CO.

The vibrator head and transformer are enclosed in two rectangular cans with a one-quarter inch thick sponge rubber between them. These cans serve as a double electrical shield and the rubber insulation serves to dampen the mechanical "buzz" to a negligible level. All parts and circuits associated with the eliminator are completely shielded from the radio set proper by the use of partitions and covers.

On earlier receivers using interrupter "B" supplies, it has been necessary to protect the vibrator during the starting period when arcing is prevalent by the use of spark gaps, relays, or other trick devices of this nature. There are definite disadvantages to all of these methods. The power supply system used in the Model 66 overcomes this difficulty in a unique fashion. A special resistance unit is shunted across the secondary of the output transformer. This has the property of changing its resistance as the voltage applied across it is varied. At a voltage of 500 volts, the resistance is approximately 500,000 ohms, while at 2500 volts, the resistance decreases to about 2500 ohms. In this manner the exceedingly high peak voltages which occur during the starting period of operation are effectively reduced, since at these voltages the unit is low in resistance. After the receiver is in an operating condition and drawing its normal load, the peak voltages are low and the unit presents a very high resistance to the flow of current through it.

REMOTE CONTROL UNIT

The airplane type control unit used on the Model 66 is unique in appearance and differs radically from earlier designs of control units. It has a full vision dial calibrated in kilocycles. Due to the use of torsional mechanical controls it is unnecessary to use a complicated cable to connect the control unit to the receiver, for with the exception of the pilot light wire, there is no electrical connection between the receiver and the control unit. The "on and off" switch and volume control are both controlled by the left hand knob while the tuning is done by the right hand knob.

BOOSTER SWITCH

On Model 66 receivers bearing serial number 16,000 and up a switch is located above the tone control which is connected across the 10,000 ohm resistor, R-9 and the .25 Mfd. Condenser, C-14. For normal reception this switch should be turned towards the left of the receiver where ample sensitivity for ordinary reception, quietest performance, maximum power and best tone will be secured.

To bring in distant or low-powered stations the switch should be turned to the right, in which position R-9 and C-4 are shorted out and an apparent increase in sensitivity will be noted.

DYNAMIC SPEAKER

A 6" full dynamic speaker, having exceptionally good response over the entire audio range, is used in conjunction with the Model 66 receiver. The speaker is adequate in size to handle with excellent fidelity, all normal output levels necessary for automobile receivers.

The receiver is equipped with twin jacks for the connection of an external magnetic type speaker, to be operated at the same time as the regular speaker. The use of the extension speaker has practically no effect upon the volume output of the regular dynamic speaker in conjunction with which it is working.

GRIGSBY-GRUNOW CO.

MOUNTING OF RECEIVER

The receiver is designed to be installed on the inside of the fire-wall behind the instrument panel, preferably in a horizontal position and close enough to the steering column for the control cables to reach the receiver. Only in cases where it is impossible to install in a horizontal position should it be mounted vertically. Mount the two adjustable brackets, one on each end of the receiver, then determine the best location for the receiver by holding it against the fire-wall, being careful to avoid interference with mechanical controls of the car. It may be necessary to reverse the brackets to accomplish this. After the best location

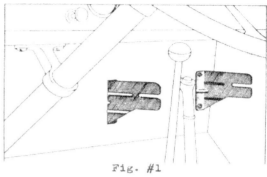

Fig. #1

has been determined, drill four holes using the template furnished with receiver for marking their location. Figure #1 shows how the brackets should look after being bolted to the fire-wall. Before permanently bolting the receiver to the brackets, the plug of the battery cable should be inserted into the rear of the receiver.

CAUTION - All mounting nuts and bolts must be drawn tight.

CONNECTING CONTROL

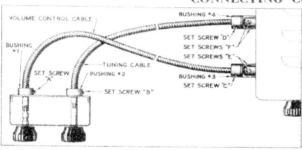

Fig. #2

Two flexible drive shafts are furnished with the Model 66 receiver. The volume control shaft has a slotted fitting on one end while the tuning shaft is similarly provided with a key fitting. To assemble the control unit the end of the volume control shaft with the slotted fitting should be inserted into bushing No. 1 on the control unit. (See Fig. #2). Make sure the outside casing of the shaft goes about five-sixteenths of an inch into the bushing. Then tighten the set screw "A" so that the outer casing of the cable will be securely held. Now connect in the same manner, the key end of the tuning cable to bushing No. 2, securing it with set screw "B". After the two cables are so connected, to sure that the knobs on the control head turn smoothly and without binding. Binding might be caused by the cables being pushed too tightly into the control unit.

The left hand or volume control cable should now be connected to bushing No. 3 on the end of the receiver. Pass the cable through the bushing so that the fitting on the end of the cable fits into the coupling on the volume control and the outer casing of the cable comes flush with the inside edge of bushing No. 3. Tighten set screw "C" so that it will securely hold the outer casing.

Next, connect in the same manner, the tuning cable to bushing No. 4, securing it with set screw "D". If the cables are properly connected they will cross. Set screws "E" and "F" should not be tightened until the control unit and cables are permanently mounted.

Now mount the control unit on the steering column in the most convenient place. Fasten drive cables securely wherever convenient so that they will not interfere with operation of the car, and then tighten the set screws "E" and "F"

GRIGSBY-GRUNOW CO.

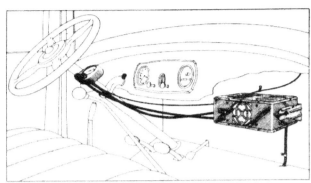

Fig. #3

in the couplings. If these are tightened before the control unit has been mounted, binding of the controls might result. Binding might also be caused by the bushings on the end of the receiver not being directly in front of the couplings. By loosening the screws that hold the bushings and then re-adjusting the bushings, this condition should be remedied. After control unit and receiver are mounted, they should appear as in Figure #3.

After the control unit and cables have been connected, the dial pointer should be adjusted. This is accomplished by slowly rotating the tuning control knob to the right until a definite stop is reached. Do not force the knob after the stop has been encountered as this may seriously damage the mechanism. Then rotate the knob slowly to the left until another definite stop is reached. In most cases it will be natural for either the pointer to come to the end of the dial strip before the stop is reached, or for the stop to be reached before the pointer comes to the end of the dial strip. In this manner the dial pointer is automatically adjusted to indicate correct frequency readings.

BATTERY CONNECTION

The shielded battery cable should now be connected directly to the car storage battery terminals, running the cable via the shortest possible route--preferably along the channel of the car chassis. In practically all cars, the terminals of the battery are badly corroded. Scrape the terminals clean so that all corrosion is removed before making the battery cable connections. Keep this cable out of the motor compartment and away from all high tension leads. Use the clamps furnished with receiver for grounding the shielding at as many points as possible. Also attempt to keep battery cable away from antenna lead-in.

IMPORTANT

The shielding of this cable must always be connected to the ground side of the battery, and the wire emerging from the shielding, to the hot side. The polarity of the battery need not be considered when making these connections. However, the receiver as shipped, is connected for installation in a car having the positive terminal of the battery grounded. If the installation is made in a car which has the negative terminal of the battery grounded it will be necessary to reverse two wires in the vibrator assembly of the receiver.

To locate these wires, first remove the top of the receiver, and then the outer vibrator assembly cover which is at the right side of the receiver. This cover is held down by four nuts. Now remove the top piece of sponge rubber and the inner vibrator cover. The vibrator armature assembly as shown in the diagram on page #12 will now be accessible. Connected to one side of the vibrator is a blue wire, and to the opposite side a black wire as indicated on the drawing. These two wires should be reversed when the receiver is installed in a car in which the negative terminal of the battery is grounded.

For positive ground:

    Wire "A" Connects to lug "A"
    Wire "B" Connects to lug "B"

For negative ground:

    Wire "A" Connects to lug "B"
    Wire "B" Connects to lug "A"

GRIGSBY-GRUNOW CO.

CONNECTING ANTENNA

Some automobiles are factory equipped with a roof antenna. In this case, the antenna should be checked to be sure it is not grounded; however, if the car is not equipped with a roof antenna, it is recommended that, in cases where it is possible, this type of antenna be installed as shown in Fig. #4. If the roof antenna cannot be used the Majestic running board antenna, part No. 8585, which can be purchased through your Majestic distributor, is recommended and

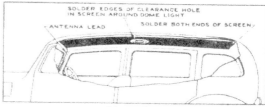

Fig. #4

is mounted as shown in Fig. #5. The single conductor shielded lead on the left hand side of the receiver is the antenna lead. This should be connected directly to the lead of the antenna. Keep lead-in out of motor compartment and away from all high tension leads and the radio set battery cable. After the connection

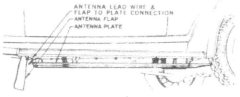

Fig. #5

to the antenna has been soldered and taped, the shield of the lead-in and the antenna lead of the set should be brought together over the taped connection and soldered to each other. Low capacity shielded antenna wire should be used if it is necessary to lengthen the lead-in. When a roof antenna is used the shielding should extend two inches beyond point where antenna lead leaves car body.

OPERATING RECEIVER

The receiver is now ready for operation and should be tested to see that all electrical connections have been properly made. The left hand knob on the control unit actuates the "On" and "Off" switch and volume control. The first fifteen degrees of this knob in a clockwise direction turns the receiver "On". Further rotation of this knob controls the volume of the receiver. The other knob on the control unit is the station selector. The knob on the lower left front corner of the receiver operates the tone control and should be adjusted according to the desire of the operator. The dial is calibrated in kilocycles so that the frequency to which the receiver is tuned may be read directly from the dial.

NOTICE - Best results from the receiver will be obtained if the car generator is set to charge at a higher rate, in order to keep the battery at full charge at all times. DO NOT, HOWEVER, SET IT TO CHARGE IN EXCESS OF 14 AMPERES.

EXTRA SPEAKER CONNECTIONS

A terminal is provided under each of two plug buttons on the left end of the receiver for the connection of an extra speaker to the receiver. This speaker is of the magnetic type and is connected to the plate of the output tube.

FUSES

For protection to the receiver and car battery, the receiver is provided with two fifteen ampere fuses. These are located in the bottom of the receiver and if replacement is necessary, may be reached by removing the bottom cover of the chassis container.

GRIGSBY-GRUNOW CO.

Every MAJESTIC Model 66 Auto Radio includes six (6) spark plug suppressors, Part No. 4640, one (1) distributor suppressor, Part No. 5122 and two (2) condensers, Part No. 8278. These accessories are to be used to prevent motor interference from being picked up by the radio receiver while the motor is running and they should be installed in the following manner.

SPARK PLUG SUPPRESSORS

Remove one at a time, the high tension lead from the top of each spark plug; mount in its place a spark plug suppressor. Connect the high tension lead to the terminal provided for it on end of the suppressor. Mount suppressor in horizontal position when possible. Figures #6 and #7 show the proper method of installing spark plug suppressors. On some cars such as the Buick, Franklin and Nash, screw type suppressors, Part No. 5199, should be used. These are installed by cutting high tension leads about two inches from the plugs. Then screw one cut end of the wire into each end of the suppressor. Be sure of a good contact. This type suppressor is shown installed in a lead in Figure #9.

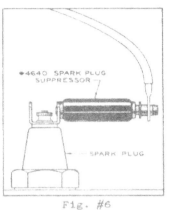

Fig. #6

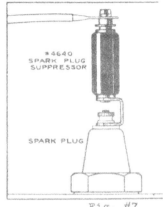

Fig. #7

DISTRIBUTOR SUPPRESSOR

Install the distributor suppressor in the center socket of distributor head, as shown in Figure #8, by removing the high tension lead which runs from the distributor head to the coil and plugging the split end of the suppressor into the distributor head, making sure of a good contact. Insert the high tension lead in the other end of the suppressor. If the car has a cap type distributor, the suppressor may be plugged in the coil or the screw type suppressor may be used, see Figure #9. In cars having two coils, a suppressor in each high tension lead is necessary. Always install the suppressor as close to the distributor as possible

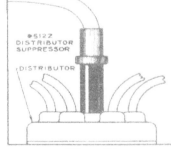

Fig. #8

*5199 SCREW-TYPE SUPPRESSOR

SCREW SUPPRESSOR INTO END OF WIRE

Fig. #9

GENERATOR CONDENSER

Fasten the lead of one of the .5 microfarad condensers to the generator side of the cut-out relay on the car generator and clamp the condenser to the frame of the generator. The screw holding the cut-out may be used for this purpose. Be sure that the condenser is securely fastened and a good ground connection made. In most cars, this condenser can be installed and connected as illustrated in Figure #10.

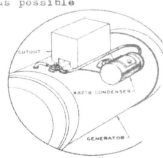

Fig. #10

GRIGSBY-GRUNOW CO.

AMMETER CONDENSER

Fasten the lead of the other .5 microfarad condenser to the storage battery
side of the ammeter. This usually is the terminal that has only one wire connected to it. Secure condenser to instrument panel (if it is metal) or to some
metal part being sure of a good ground connection.
A typical installation of this condenser is shown
in Figure #11. Sometimes this condenser is more
effective when attached to the dome light, stop
light or horn wires. The latter is usually necessary when the car is equipped with a roof antenna.
This may be tried while the motor is running and
the effect on the interference noted; the condenser
being connected to most effective point. It may be
necessary in extreme cases to connect a condenser
to more than one of these points in order to obtain reception free from interference.

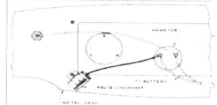

Fig. #11

In some cases, if after installing the suppressors and condensers as directed
and motor noise still exists, shielding of the pilot light wire that runs
from the receiver to the control head will very effectively suppress the disturbance.

The above procedure should effectively suppress motor interference in practically all installations. However, if this does not hold true, it may be necessary
to apply one or more of the following methods before complete suppression is
obtained.

MISCELLANEOUS

First, determine whether the radiation is picked up by the antenna or by the
receiver itself. This can be done by grounding the antenna lead as it leaves
the receiver. If the motor interference stops, one may be sure that is is
being picked up by the antenna. If it continues it is quite certain that part
of the noise is being picked up by the receiver itself. If this is the case,
make sure that all ground connections are clean and tight. If the instructions
for installing the receiver have been carefully followed and all wires of the
radio set have been kept out of the motor compartment, there should be no receiver pick-up. In the event of antenna pick-up of motor interference, the
following suggestions are made to eliminate it. These suggestions should be
followed in the order in which they are given and the motor started and tested
for interference after each step.

It may be necessary to reduce the gap between the rotor arm and contacts of the
distributor head. Extreme care should be used in this operation to prevent
harming the distributor. Peen the rotor by placing it on a flat steel block
and hammering the end with a small machinist's hammer. Repeat this operation
until there is just sufficient clearance - about .004". The rotor must not be
allowed to touch the contacts. If there is evidence of the rotor touching the
contacts, file off about .001" and recheck. Building up the rotor arm with
solder is not recommended as the solder is very soon burned away. In some
cases, where the rotor is badly worn, it may be best to substitute a new one.

GRIGSBY-GRUNOW CO.

If the motor interference still continues, it may be well to determine the source. This can be done by removing the high tension lead from the coil to the distributor, turning on the ignition switch and cranking the car by hand. If a clicking is heard in the speaker, you may be sure that part of the trouble comes from the breaker points in the distributor or low tension circuit. It will then be necessary to remove the primary lead which runs from the coil to breaker points on the distributor, and replace it with a No. 14 shielded low tension cable, being sure not to run close to the high tension leads. The shielding must be grounded in at least two places. All ground connections must be as short as possible. It may be necessary to remove the lead from the switch to the coil and replace with a No. 14 shielded low tension cable being sure to ground the shielding. Care must be used when shielding so as not to short the coil or switch. Never use a by-pass condenser on this part of the circuit because it will effect the operation of the motor.

When you have tested to determine the source of motor interference and no clicking was heard in the speaker, we may assume that the interference is coming from the high tension or secondary circuit which is possibly the worst source of motor interference. All wires which run parallel to or within the field of this part of the circuit act as carriers and should be moved whenever possible, or the high tension wire re-routed. Sometimes the car manufacturer utilizes the high tension manifold to hold various wires and just removing them from the manifold will be sufficient. Be careful to keep the high tension lead as far as possible from the receiver. If after moving the wires, the interference continues, the high tension lead should be shielded. Care should be used when shielding the high tension lead to prevent the current from leaking through to ground. To prevent this first cover the high tension lead with loom, then run this shielding over the loom. The shielding must be grounded in at least two places (to the coil and motor block or high tension manifold). When the coil is under the cowl or bulkhead, the high tension lead should run as direct as possible to the motor compartment. This will sometimes necessitate drilling a new hold about one-half inch in diameter in the firewall or dash.

Due to the electro-magnetic field surrounding the ignition coil, it may be necessary, when the coil is under the cowl or bulkhead, to move it into the motor compartment. Mount it on the motor block as close to the distributor as possible and be sure that a good ground connection is maintained. If it is found necessary to mount the coil over the motor, care should be taken that it is so mounted as to stay sufficiently cool. New primary wires will be required and shielded No. 14 low tension ignition cable should be used. Caution! Do not run these wires close to the high tension lead, but ground them well. ONLY MOVE THIS COIL AS A LAST RESORT.

In a number of cases, the establishing of a good electrical contact between the motor block, firewall and frame of the car will eliminate much of the interference. In assembling automobiles, often times paint or other substances will prevent a good ground connection from being made between the various metal parts of the car which form the ground circuit. These poor connections will have no apparent effect on the operation of the car. However, when a radio receiver is installed, it is especially desirable to maintain all the metal parts of the car at the same ground potential. This is accomplished by connecting together with short pieces of shielding the motor block, frame, and firewall and sometimes the body of the car. Bonding may be particularly necessary on those cars having the motor mounted on rubber blocks. When bonding the motor to the firewall, use one inch shielding and make the bond long enough to allow for vibration of the motor.

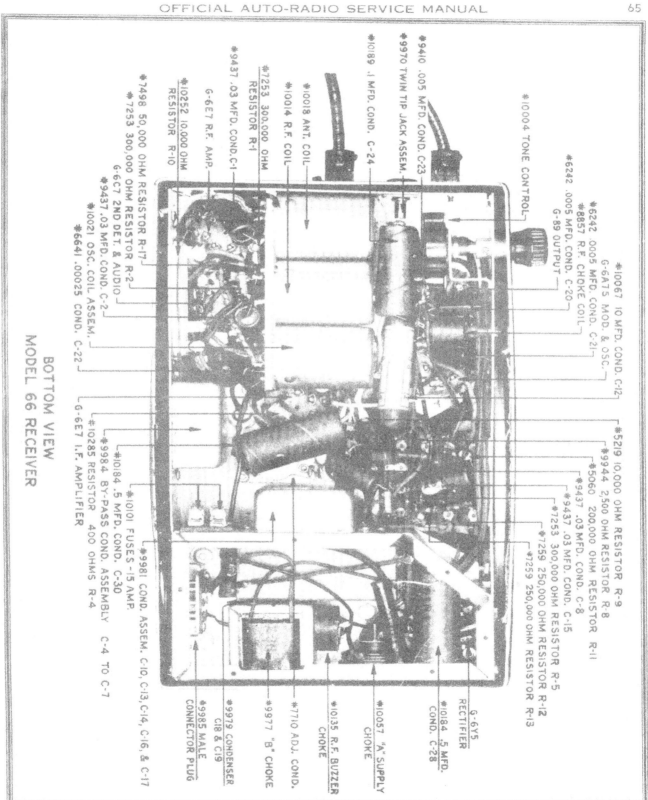

BOTTOM VIEW
MODEL 66 RECEIVER

*9410 .005 MFD. COND. C-23
*9970 TWIN TIP JACK ASSEM.
*10189 .1 MFD. COND. C-24
*6242 .0005 MFD. COND. C-21
*8857 R.F. CHOKE COIL
*6242 .0005 MFD. COND. C-20
G-89 OUTPUT
G-6A75 MOD. & OSC.
*10067 10 MFD. COND. C-12
*10004 TONE CONTROL

*9437 .03 MFD. COND. C-1
*7253 300,000 OHM RESISTOR R-1
*10018 ANT. COIL
*10014 R.F. COIL

*7498 50,000 OHM RESISTOR R-17
*7253 300,000 OHM RESISTOR R-2
G-6C7 2ND DET. & AUDIO
*9437 .03 MFD. COND. C-2
*10021 OSC. COIL ASSEM.
*6641 .00025 COND. C-22

*10252 10,000 OHM RESISTOR R-10
G-6E7 R.F. AMP.

*5219 10,000 OHM RESISTOR R-9
*9944 2,500 OHM RESISTOR R-8
*5060 200,000 OHM RESISTOR R-11
*9437 .03 MFD. COND. C-8
*9437 .03 MFD. COND. C-15
*7253 300,000 OHM RESISTOR R-5
*7259 250,000 OHM RESISTOR R-12
*7259 250,000 OHM RESISTOR R-13

*10184 .5 MFD. COND. C-28
G-6Y5 RECTIFIER
*10057 "A" SUPPLY CHOKE
*10135 R.F. BUZZER CHOKE
*7710 ADJ. COND.
*9977 "B" CHOKE
*9979 CONDENSER C18 & C19
*9985 MALE CONNECTOR PLUG
*9984 BY-PASS COND. ASSEMBLY C-4 TO C-7
*10285 RESISTOR 400 OHMS R-4
*10101 FUSES -15 AMP.
*9981 COND. ASSEM. C-10, C-13, C-14, C-16, & C-17
G-6E7 I.F. AMPLIFIER

GRIGSBY-GRUNOW CO.

Each and every wire, rod or pipe that runs from the motor compartment through the firewall into the body of the car acts as an antenna to radiate interference and should not be overlooked. To stop them from radiating, solder a heavy flexible copper conductor to them close to the firewall, allowing room for any movement of the rods, and then ground each of these to the firewall. If they are rusty, scrape them clean where contact is made. The wire conduit that runs to the base of the distributor on some cars should be grounded the same as other wires or rods.

In some instances, the noise being heard in the receiver will be caused by loose wires at the headlights, horn or horn button, tail light, spot light, stop light or dome lights. All connections to these items should be checked to see that the contact surfaces are clean and all wire connections are tight. Sometimes the connecting of a Part #8278, .5 microfarad condenser to the hot battery lead feeding one or more of these accessories will have a decided effect on the ignition noise. This is especially true of the dome light wires when a roof type antenna is being used.

Any metal parts about the set making imperfect or intermittent contact with each other will cause noise in the speaker when the car is subjected to a jolt, whether there is any measurable potential difference between these parts or not. This interference is due to the instantaneous change in resistance of the receiver to ground that occurs when another ground conductor touches or is disconnected from the receiver.

To guard against the possibility of such noises, choke wires, speedometer cables, copper tubes, battery cables or the like should not be allowed to rub on the radio container. Also, always make sure that the chassis mounting bolts and the bolts that hold the brackets are securely tightened so that there will be no possibility of the contact resistance to ground changing.

If the foregoing information is insufficient to give suppression of motor interference, write in detail to the Majestic distributor in your territory or to the Grigsby-Grunow Company.

VOLTAGE TABLE FOR MODEL 66 AUTO RECEIVER

Battery) Volts Terminal)	PLATE VOLTS			SCREEN VOLTS			CATHODE VOLTS			GRID VOLTS		
	5.5	6.3	7.5	5.5	6.3	7.5	5.5	6.3	7.5	5.5	6.3	7.5
R. F. (G-6E7)	182	217	256	88	99	109	8.0	9.3	12.5	8.0	9.3	12.5
G-6A7S Det.	182	217	256	88	99	109	2.7	3.4	4.2	2.7	3.4	4.2
Osc.	88	99	109	-	-	-	-	-	-	7.0*	8.0*	8.0*
I. F. (G-6E7)	182	217	256	88	99	109	8.0	9.3	12.5	8.0	9.3	12.5
Audio (G-6C7)	51	60	61	-	-	-	7.5	9.2	9.5	1.8	2.2	2.3
Output (G-89)	177	209	248	184	218	257	-	-	-	23.0	27.0	35.0

Battery Terminal Volts	5.5	6.3	7.5	* Measured with 300,000 ohm meter.
B+ to B- (Volts)		216	261	322
B+ to Ground (Volts)		184	218	257
Total Battery Drain (Amps)		6.15	7.25	8.50

B+ to B- (Volts)   216   261   322   All voltages measured with no input signal

B+ to Ground (Volts)   184   218   257   All voltages to ground from socket unless

Total Battery Drain (Amps)   6.15   7.25   8.50   otherwise stated.

GRIGSBY-GRUNOW CO.

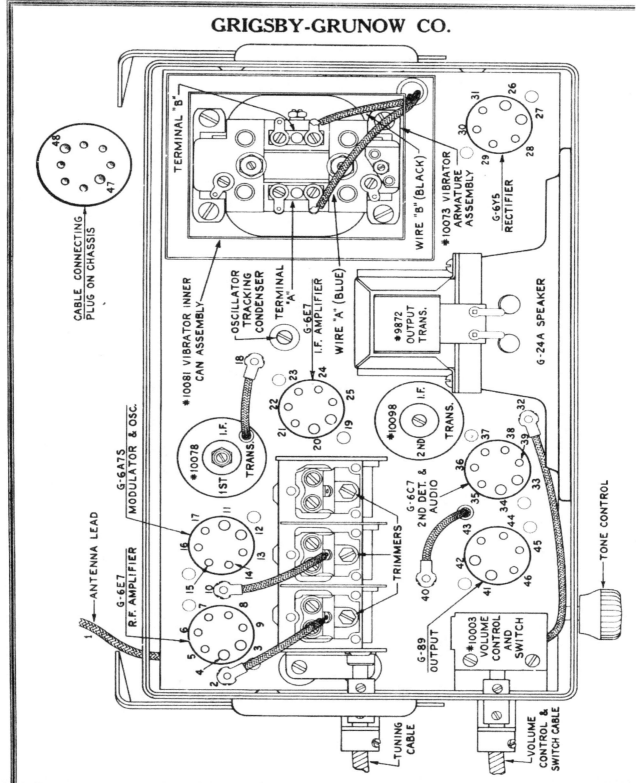

GRIGSBY - GRUNOW CO.

Resistors

R1— 300,000
R2— 250
R3— 300,000
R4— 400
R5— 300,000
R6— 100,000
R7— 200,000
R8— 6,500
R9— 10,000
R10— 10,000
R11— 200,000
R12— 250,000
R13— 250,000
R14— 50,000
R15— 300,000
R16— 500,000 (GLOBAR)
R17— 500,000
R18— 1,000,000

Condensers

C1— .03
C2— .03
C3— .01
C4— .1
C5— .25
C6— .25
C7— .25
C8— .03
C9— .0005
C10— .03
C11— .0005
C12— 10.
C13— .25
C14— .25
C15— .03
C16— .25
C17— .02
C18— 8.0
C19— 8.0
C20— .0005
C21— .0005
C22— .00025
C23— .005
C24— .1
C25— .008
C26— .008
C27— .1
C28— .5
C29— .1
C30— .5

◆ Note ◆

When A+ is Grounded Vibrator Lead #1 (Blue) should connect to Terminal "M" (Vibrator Armature) and Lead #2 (Black) should connect to Trans. Primary Center Tap (Terminal "N") When A- is Grounded Reverse Above Connections.

Grigsby-Grunow Co.
Chgo, Ill.

Schematic Diagram of Majestic Model 66 Automobile Receiver.

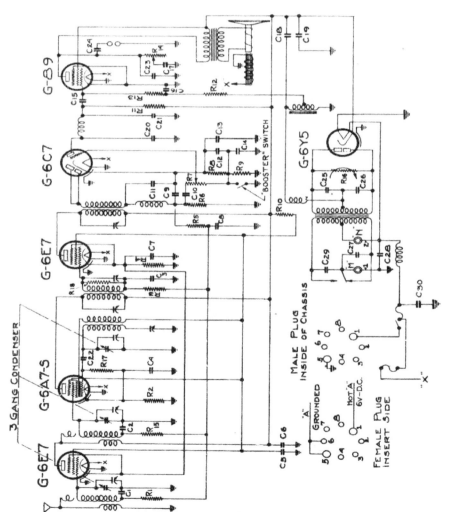

3 Gang Condenser

G-6E7   G-6A7-S   G-6E7   G-6C7   G-89   G-6Y5

"Booster" Switch

Male Plug Inside of Chassis

Female Plug Insert Side

Grounded

Hot "A" 6v-DC.

MODEL 66 RESISTANCE CHART

All readings are taken from designated points to ground except those marked with an asterisk (*) which are taken to terminal No. 29, with all tubes removed from their sockets, volume control turned to maximum clockwise position, and the speaker connected in the circuit.

TERMINAL NUMBER	RESISTANCE IN OHMS	IF RESISTANCE DIFFERS GREATLY FROM VALUE SHOWN, CHECK THE FOLLOWING:
1	21	Primary of antenna coil
2	700,000	Secondary of antenna coil, R-1, C-1, R-5, C-8 and R-6
3	0	Ground connection
4	.135	Primary of vibrator trans., Field Coil, C-30, C-28, C-27 and C-29
5	400	R-4 and C-7
6	0	Ground connection
7	Same as #5	
*8	10,000	R-10
9	112	Primary of R.F. transformer
10	700,000	Secondary of R.F. transformer, C-2 and R-15
11	Same as #4	
12	0	Ground connection
13	250	R-2 and C-4
14	50,250	R-17
*15	10,000	Secondary of oscillator coil and R-10
16	Same as #8	
*17	88	Primary of 1st I.F. transformer
18	700,000	Secondary of 1st I.F. transformer, C-3, and R-3
19	Same as #4	
20	0	Ground connection
21	Same as #5	
22	0	Ground connection
23	Same as #5	
24	Same as #8	
*25	165	Primary of 2nd I.F. transformer
26	Same as #4	
27	0	Ground connection
28	1250	Secondary of vibrator trans., C-26, C-25, R.F. buzzer choke, and "B" filter choke
29	0	C-18, C-19, C-5 and C-6
30	Same as #28	
31	0	Ground connection
32	210,000	C-10, R-7, R-9, C-14 and C-13
33	Same as #4	
34	0	Ground connection
35	12,500	R-8, R-9, C-12, C-13, C-14 and C-10
36	100,284	Secondary of 2nd I.F. trans., R.F.C., R-6, C-11, C-9 and C-10
37	Same as #36	
38	0	Ground connection
*39	200,035	C-20, C-21, R.F.C., C-15 and R-11
40	500,450	R-13, R-12, C-16 and "B" filter choke
41	Same as #4	
42	0	Ground connection
43	0	Ground connection
44	Same as #43	
*45	0	Connections
46	430	Primary of output transformer
47	0	Ground connection
48	Same as #4	

GRIGSBY - GRUNOW CO.

Due to manufacturing tolerances on carbon resistors, the values given above may be expected to differ plus or minus 15 per cent.

GRIGSBY-GRUNOW CO.

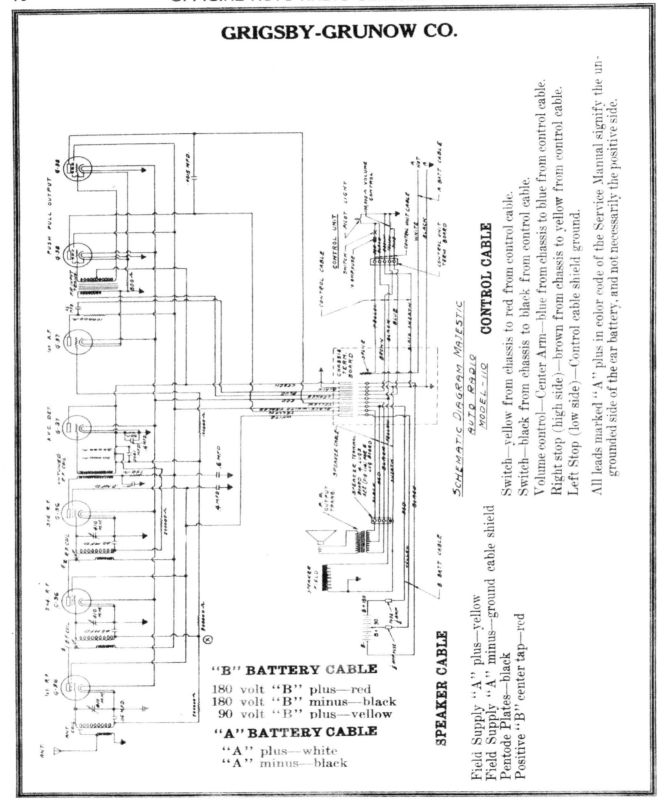

SCHEMATIC DIAGRAM MAJESTIC
AUTO RADIO
MODEL-110

CONTROL CABLE

Switch—yellow from chassis to red from control cable.
Switch—black from chassis to black from control cable.
Volume control—Center Arm—blue from chassis to blue from control cable.
Right stop (high side)—brown from chassis to yellow from control cable.
Left Stop (low side)—Control cable shield ground.

All leads marked "A" plus in color code of the Service Manual signify the un-grounded side of the car battery, and not necessarily the positive side.

"B" BATTERY CABLE

180 volt "B" plus—red
180 volt "B" minus—black
90 volt "B" plus—yellow

"A" BATTERY CABLE

"A" plus—white
"A" minus—black

SPEAKER CABLE

Field Supply "A" plus—yellow
Field Supply "A" minus—ground cable shield
Pentode Plates—black
Positive "B" center tap—red

GRIGSBY - GRUNOW CO.

CONTINUITY AND RESISTANCE CHECK OF SPEAKER FOR MODEL 114 RECEIVER

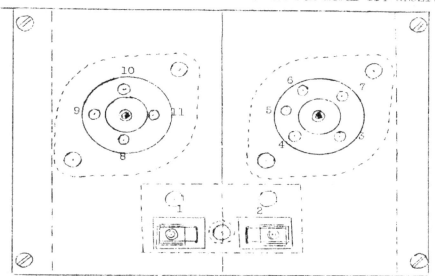

Note - Checks made with fuse in place. All plugs removed.

Terminal Number	Resistance	If results differ greatly from value shown, check the following
12,5,6 and 8	All short circuit to each other	Wiring between points
13 to 4	Short circuit	Fuse
3 to 11	Short circuit	Wiring
7 to 9	Short circuit	Wiring
*12 to 13	Open	Short in "A" cable
10 to 8	1-1/2 ohms	Speaker cone voice coil
9 to 8	6 ohms	Speaker field coil

CONTINUITY AND RESISTANCE CHECK OF CONTROL UNIT FOR MODEL 114 RECEIVER

Terminal Number	Resistance in Ohms	If results differ greatly from value shown, check the following
1 to 3	0 to 250,000	R9, volume control
3 to 4	Short circuit	Wiring in plug
2 to 5	Short switch "on" Open switch "off"	On and off switch
2 to 3	Low resistance	pilot bulb

NOTE: * #12 Shielded lead of
battery cable.

#13 White lead of battery
cable

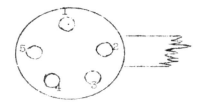

Control Unit Plug

GRIGSBY - GRUNOW CO.

MODEL 114 RECEIVER CHASSIS (Cont'd)

- 2 -

Terminal Number	Resistance in Ohms	If results differ greatly from value shown, check the following
41 to 35	Short circuit	Wiring between points
35 to chassis	Open	C-1, C-3, C-7
38 to chassis	Same as #35 above	
42 to 43	7,400	Secondary of input transformer
42 or 43 to chassis	3,700	Secondary of input transformer or center tap to ground
34 to chassis	Low resistance	Output transformer secondary
20 to chassis	Open	C-6, C-5, C-4

If the above tests show a normal condition, it will be necessary to check the units and wiring in the circuits where the difficulty has been previously localized.

TUBE	PURPOSE IN CIRCUIT	PLATE VOLTAGE	SCREEN VOLTAGE	CATHODE VOLTAGE
G-39	R. F. Amplifier	180	85	0
G-38	1st Detector Oscillator	180	85	15.
G-39	I.F. Amplifier	180	85	1.1
G-85	2nd Detector and 1st Audio Amplifier	A.F.Plate 50	--	2
G-38	Power Amplifier	170	180	17
G-38	Power Amplifier	170	180	17

GRIGSBY - GRUNOW CO.

CONTINUITY AND RESISTANCE CHECK
MODEL 114 RECEIVER CHASSIS

NOTE: Readings taken with all tubes and cable plugs removed from their sockets.

Before making the following tests, check for shorted C-13, gang condenser or I.F. trimmers.

Due to manufacturing tolerances on carbon resistors, the values given below may be expected to vary plus or minus 15 per cent.

* Note-that the readings vary according to the polarity of the test leads due to the presence of electrolytic condensers.

Terminal Number	Resistance in Ohms	If results differ greatly from value shown, check the following
Ant. to chassis	Approximately 25	L-1 or connections thereto
2 to chassis	Short circuit	Solder connection #2 to chassis
7 to chassis	8,000	L-7, R-2, C-2
12 to chassis	150	R-10
*17 to chassis	2,000	R-5, C-9
23 to chassis	1,000	R-8
28 to chassis	Same as #23	
3, 8 and 13	Short circuit to each other	Wiring between points
3 to chassis	Open	C-4, C5
3 to 36	30,000	R-4
4 to 9	Approximately 120	L-3, L-8
9 to 14	Approximately 205	L-8, L-10
14 to 20	99,140	L-10, R-7
14 to 24 and 29	Approximately 140	L-10
4 to 36	Approximately 55	L-3
4 to chassis	Open	C-5, C-4
25 to 30	Approximately 1,600	C-8 or output transformer primary
25 or 30 to 36	Approximately 800	Output transformer primary or center tap to 36
5,10,16,21,27,32,33	All short circuit to each other and open to ground	
6,11,15,22,26,31,37	All short circuit to chassis & each other	Wiring between points
18 to 19	Short circuit	Ground connections to sockets
39 to chassis	Approximately 8	Wiring between points
38 to 40	99,085	L-4
40 to 18	500,225	R-1, L-9 or C-1, C3
18 to 41	99,140	R-6,L-9, L-11 or C-7, C-3; L-11, R-3

GRIGSBY - GRUNOW CO.

GRIGSBY-GRUNOW CO.

CONTINUITY of 3 PLUG CABLE FOR MODEL 114 RECEIVER

CHASSIS - PLUG   GENERATOR - PLUG   SPEAKER - PLUG

NOTE—For intermittent opens or shorts, cable should be strained and shaken while continuity tests are being made.

Terminal Numbers	Result	
1 to 13	Short circuit	
2 to 7 to 10	Short circuit	Check all solder connections
3 to 6 to 11 to shield	Short circuit	To plug contacts
4 to 9	Short circuit	If okay then wiring within cable is defective, replace entire assembly
5 to 12	Short circuit	
8	Not used	

All contacts except 3, 6 and 11 are open to shield.

GRIGSBY - GRUNOW CO.

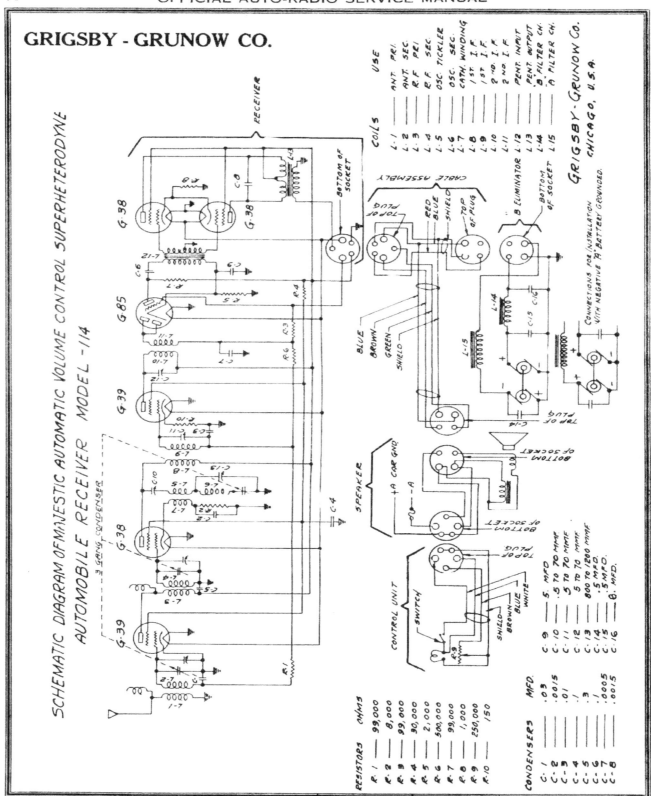

SCHEMATIC DIAGRAM OF MAJESTIC AUTOMATIC AUTOMATIC VOLUME CONTROL SUPERHETERODYNE
AUTOMOBILE RECEIVER MODEL -114

GRIGSBY - GRUNOW CO.
CHICAGO, U.S.A.

COILS   **USE**

L-1	ANT. PRI.
L-2	ANT. SEC.
L-3	R.F. PRI.
L-4	R.F. SEC.
L-5	OSC. TICKLER
L-6	OSC. SEC.
L-7	CATH. WINDING
L-8	1st I.F.
L-9	1st I.F.
L-10	2nd I.F.
L-11	2nd I.F.
L-12	PENT. INPUT
L-13	PENT. OUTPUT
L-14	"B" FILTER CH.
L-15	"A" FILTER CH.

CONNECTIONS FOR INSTALLATION
WITH NEGATIVE "A" BATTERY GROUNDED.

RESISTORS   **OHMS**

R-1	99,000
R-2	8,000
R-3	99,000
R-4	30,000
R-5	2,000
R-6	500,000
R-7	99,000
R-8	1,000
R-9	250,000
R-10	150

CONDENSERS   **MFD.**

C-1	.03
C-2	.0015
C-3	.01
C-4	.3
C-5	.1
C-6	.0005
C-7	.0015
C-8	
C-9	5. MFD
C-10	.5 TO 70 MMFD.
C-11	.5 TO 70 MMFD.
C-12	5 TO 70 MMF.
C-13	800 TO 1200 MMFD.
C-14	.5 MFD.
C-15	.5 MFD.
C-16	8. MFD.

GRIGSBY - GRUNOW CO.

MAJESTIC MODEL 116 AUTO RADIO

INTRODUCTION

THE CIRCUIT - The MAJESTIC Model 116 is the very latest in auto radios, being a six tube superheterodyne giving the performance of an eight tube receiver due to the dual operation of two of the tubes.   The tubes used in the various stages are as follows: First detector and oscillator, G-57A-S; first intermediate frequency amplifier, G-58A-S; second intermediate frequency amplifier, G-58A-S; second detector and first audio frequency amplifier, G-75; output audio amplifier, G-89; and full-wave rectifier, G-6Z5.   For protection to the radio receiver and the car battery against any possible damage due to shorts or grounds, both the primary circuit of the "B" supply, and the tube filaments and field coil circuit have a ten ampere fuse in series with them.   These two fuses are located on the control unit.

The receiver is characterized by its high sensitivity and it will be found that on local reception, only a small degree of increase in volume will be sufficient to obtain normal reception in the car.

AUTOMATIC VOLUME CONTROL - The Model 116 MAJESTIC Auto Radio is equipped with an efficient automatic volume control system which does away with blasting and fading.

This is accomplished by the space current drop of the G-75 diode detector plate circuit across a resistor, the negative potential of which is applied to the grid of both the in-termediate frequency amplifiers to control their amplification.

By having the receiver equipped with an efficient Automatic Volume Control system, the re-ception will not completely disappear when passing under a grade crossing or over a steel bridge, nor will it gradually diminish when traveling away from the station broadcasting the program being received.   This makes it unnecessary to continuously adjust the volume control so that the output of the receiver will remain at the same level.

THE DURO-MUTE POWER SUPPLY - The Duro-Mute Power Unit, which is employed in the MAJESTIC Model 116 Automobile Receiver is composed of a transformer and a vibrator assembly unit which supplies the high voltage necessary for the efficient operation of the receiver.   It is completely housed in a metal container to attain efficient shielding and the container cover is lined with sponge rubber to insure quiet operation.

This newly designed Duro-Mute Power Unit is the first assembly of its kind to be housed in the same container with the chassis itself.

The efficiency of this assembly is increased tremendously inasmuch as the vibrator assembly unit itself is mounted directly over the transformer and receives its magnetic impulses from the core of the transformer making it unnecessary to provide additional excitation for the vibrator, thereby increasing its percentage of efficiency to a high degree.

The vibrator is so designed that when connecting to the battery, it is not necessary to consider the polarity of the battery of the car in which the radio is to be installed.

A special rectifier tube is employed in the rectifier system of the MAJESTIC Duro-Mute Power Unit, and is of the 6.3 volt heater type, and has been designed to withstand the abuse to which tubes in Automotive Receivers are subjected.

THE DYNAMIC SPEAKER - The MAJESTIC Model 116 Auto Radio is equipped with a dynamic speaker capable of handling all the volume that could be desired in an automobile receiver.   It is well constructed and will give faithful reproduction over a wide range of the musical scale without distortion.

EASY TO INSTALL - Due to the single unit construction of the receiver chassis, dynamic speaker and Duro-Mute Power Supply. the installation is very simple and requires only a

GRIGSBY-GRUNOW CO.

short time to install in any type of car. To install the radio, it is only necessary to
drill four small holes, two for each bracket on which the receiver is suspended, bolt the
receiver to these brackets, clamp the control unit to the steering column, connect the
battery cable to the car battery and connect the antenna to the receiver. It is then ready
for operation.

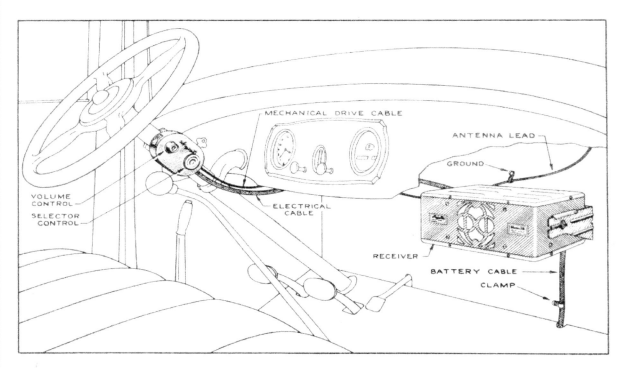

Fig. 1

Due to the simplicity of installation of the MAJESTIC Model 116 Auto Radio, only a few
tools are necessary, such as screw driver, drill, wrench and pliers. The receiver is also
quickly installed, due to the fact, that it is only necessary to mount one unit, unlike
other receivers wherein the radio chassis, power supply and speaker are all separate units
and have to be individually mounted. The installation of the MAJESTIC Model 116 Auto Radio
is effected almost as quickly and easily as that of a radio in the home. Fig. 1 shows the
simplicity of a complete installation.

EASY TO CONTROL – After the receiver is installed, it is easy to operate. All the necessary
controls are located on the control unit which is easily within the reach of the driver or
other occupants of the front seat of the car. The set is turned "on" by means of a small
key which may be removed, if desired, after turning "off" the receiver, thus acting as an
effective lock for the set and preventing unauthorized use. The volume control and tuning
control knobs are located on the side of this unit, therefore, it is not necessary to twist
the wrist to an awkward position as would be necessary if they were mounted on the upper
part of the control unit.

GRIGSBY - GRUNOW CO.

INSTRUCTIONS FOR INSTALLING

MOUNTING RECEIVER — The receiver must be installed on the inside of fire wall behind the instrument panel either in a horizontal or vertical position. Mount the two brackets, which are adjustable, one on each end of the receiver, then determine the best location for receiver by holding it against the fire wall, being careful to avoid interference with mechanical controls of the car. It may be necessary to reverse the brackets to accomplish this. After the location has been determined, drill four holes using the template furnished with the receiver. The center to center dimension of these holes should be 10 15/16 inches when the flanges of the brackets are pointing "in" and 13 3/16 inches when the flanges are pointing "out". Figure 2 shows how the brackets should look after being bolted to the firewall. Next, place the receiver on the brackets and securely tighten the bolts at each end. Always make sure that the large flat washers are under the nuts on both the bracket mounting bolts and the chassis mounting bolts.

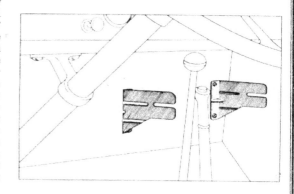

Fig. 2

MOUNTING CONTROL — First be sure that the fuses on the control unit are securely in place. Mount control unit on steering column as shown in Fig. 1 using the two clamps, four screws and lockwashers furnished for that purpose. Make sure there are no sharp bends or kinks in the mechanical cable. The leather shims furnished with the receiver are to be used when the steering column is under standard size. After the clamps are drawn up tight, screw down the set screws in the center of the clamps so that they make a firm contact in the metal of the steering column. Fasten the cables securely to prevent interference with any mechanical controls of the car.

BATTERY CONNECTION — The two conductor shielded cable at the right hand side of the receiver must be connected directly to the car storage battery terminals using shortest possible route – preferably the channel of the car chassis. Keep this cable out of the motor compartment and away from all high tension leads using the clamps furnished with the receiver for grounding the shielding at as many points as possible. The shielding is to be connected to the grounded side of the battery and the two wires emerging from the shielding are both connected to the hot side. The polarity of the battery need not be considered when making these connections. When making the ground connections, scrape away any corrosion, paint or rust so as to make a good electrical contact. TO OBTAIN BEST RESULTS FROM THIS RECEIVER, ADVANCE THE CAR GENERATOR TO KEEP THE STORAGE BATTERY FULLY CHARGED.

The cable must be securely clamped and must not come in contact with the battery in order to avoid the possibility of corrosion and shorting of the battery.

ANTENNA — Some automobiles are factory equipped with a roof antenna. This may be used very effectively, however, it should be checked to make sure that it is not grounded in any way. A test for ground can easily be made by touching the antenna lead-in to the hot side of the storage battery. If it does not spark, the antenna is not grounded and is okeh to be used.

In cases where this antenna is grounded and the ground point cannot be located, or the car is not equipped with an antenna, one has a choice of installing two different types of antennae; namely, the running board antenna or the roof antenna.

Running Board Plate Antenna — For ease of installation and minimum requirement of time the Majestic #8585 Running Board Antenna is recommended and may be purchased from the Majestic Distributor in your territory. The Majestic Running Board Antenna comes complete with instructions for mounting and is shown completely installed in Figure 3.

GRIGSBY-GRUNOW CO.

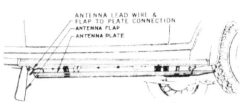

Mounting of Running Board Antenna

Fig. 3

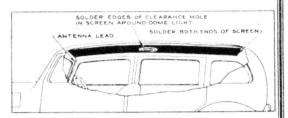

Fig. 4

Roof Antenna - For efficiency and best results for distance reception, a Roof Antenna should be used. There are four types of top construction commonly used by the automobile manufacturer. First, tops with slat construction: In these cars the headlining should be lowered, working from the front to rear. This can be done by removing the moulding between the windshield and the top of the car which is usually held in place by two or three screws. Then removing the moulding on both sides that runs from the front of the car to the back of the rear door. When this is removed you will notice the headlining is tacked to the trim rail. Remove the tacks from this, and the headlining will drop down. When replacing this headlining if care is taken to put the tacks back in their original holes, and moulding put back in place, it will be hard to tell that it has been taken down. After the headlining is down, if the top is of slat construction, #18 rubber covered stranded wire may be strung back and forth between the slats, tacking it to the front of the top and to the last bow used. About 60 to 75 feet of wire is sufficient. Be sure to keep the wire at least four inches from the metal sides of the top which is called the quarter deck. A lead-in should be fastened to one end of this wire and brought down through the corner post most convenient to the location of the receiver. It is also possible to use, instead of this stranded wire, copper screening. When this latter is used, care should be taken that the screen is kept at least three inches from any metal part of the car and the dome light. See Figure 4 which shows how this type of antenna should be installed. A stranded copper, rubber and cotton covered lead-in wire should then be soldered to the front corner nearest the receiver and then run down through the corner post. Be sure that the screen is tacked securely to the bows, being careful not to tack the screen to those bows to which the headlining strip is fastened.

Tops With Wire Construction - The headlining is removed by following the same procedure as above. The wire mesh may be used as an antenna by cutting out a three inch strip around the four sides. The center portion of the mesh is then laced securely to the part still remaining attached to the car by use of a strong cord. This should be pulled tight enough to hold the center portion of the mesh up and to prevent the top from sagging. A lead-in should be soldered to the corner of this mesh nearest the receiver and run down the corner post. The dome light wires may have to be re-arranged so that there is a minimum of coupling between them and the antenna.

Fabric Top Construction - The same procedure can be followed as in the slat top construction with the exception that if you use a copper screen, it should be placed on top of the bows and tacked at both ends.

Cars with Metal Braces - Some cars have metal diagonal braces to strengthen the top and usually these braces are fastened in wood at the rear and in a metal frame at the front. It will be necessary that these braces are freed of grounds or the efficiency of the antenna will be greatly reduced. This can be done by removing the braces at the front and reaming the holes to allow the use of a fibre washer or sleeve bushing to insulate the cross brace bolts from the brackets. Usually one of the dome light wires is connected to one of the braces and this lead will have to be disconnected from the brace and a new lead run to the body of the car.

Roadster or Convertible Type Tops - A wire antenna is the only practical one for this type of roof. Remove the tacks from the front end of the top and lay the top back exposing the bows and quarter deck pads. Use a #18 stranded rubber covered copper wire (about 75 feet)

GRIGSBY-GRUNOW CO.

starting at one corner and tacking securely in place: Then sew the wire to the quarter deck fabric as far as you can go on the same side. Then return and repeat this operation until there are six or seven rows of wire sewn into the fabric about one inch apart. Then tack wire across top box to the other side and repeat operation. The stitches are really individual knots about five or six inches apart. The lead-in should be run down the back corner which is most convenient to receiver so as to permit the folding back of car top. Care should be exercised when replacing top to get tacks in the same holes.

The proper connection of the antenna to the receiver is very important and should be made with a low-capacity shielded single conductor wire. One end should be connected directly to the shielded single conductor lead on the left hand side of the receiver. After the electrical connection has been soldered and taped, the two shieldings should be telescoped (one run over the other) for about an inch or two and soldered together. Next, a large sized wire or piece of shielding should be soldered to the shielding of the antenna lead where it leaves the chassis and to the lug on the right rear side of the chassis container and then grounded to the firewall. All paint and rust should be scraped away before the ground is made and if a bolt is used it should be tightened as much as possible. The antenna lead should now be run to the antenna, keeping it out of the motor compartment and well away from any high tension leads or the coil, and connect it to the antenna making sure of a good clean contact. If the running board antenna is used, the shielding should stop about 6" from the plate of the antenna, and if a roof antenna is used, it is only necessary to shield the lead-in to a point about two inches beyond where the lead-in enters the door post or windshield post. If the lead-in to the roof antenna is not protected by some body post, the shielding should run to within 6" of the antenna.

If the shielding of the antenna lead is grounded in two or three places it may help greatly in obtaining reception free from motor noises.

OPERATION - The receiver is now ready for operation and should be tested to see that all electrical connections have been properly made by turning the key switch located on the control unit. It will be necessary to wait approximately one-half minute before being able to obtain reception. After this length of time has elapsed, turn the small volume control knob clockwise as far as it will go, then tune in a station by rotating the large knob until a signal is heard. Adjust the small control knob until the desired volume is obtained.

NOISE SUPPRESSION

MOTOR INTERFERENCE SUPPRESSION - Every MAJESTIC Model 116 Auto Radio includes six (6) spark plug suppressors, part No. 4640, one (1) distributor suppressor, Part No. 5122 and two (2) condensers, Part No. 8278. These accessories are to be used to prevent motor interference from being picked up by the radio receiver while the motor is running and they should be installed in the following manner.

SPARK PLUG SUPPRESSORS - Remove one at a time, the high tension lead from the top of each spark plug; mount in its place a spark plug suppressor, and connect the high tension lead to the terminal provided for it on end of the suppressor. Mount suppressor in horizontal position when possible. Figures 5 and 6 show the proper method of installing spark plug suppressors. On some cars such as the Buick, Franklin and Nash, screw type suppressors, Part No. 5199, should be used. These are installed by cutting high tension leads about two inches from the plugs. Then screw one cut end of the wire

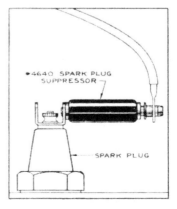

Fig. 5

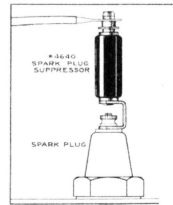

Fig. 6

GRIGSBY - GRUNOW CO.

into each end of the suppressor being sure of a good contact. This type suppressor is shown installed in a lead in Figure 7.

<u>DISTRIBUTOR SUPPRESSOR</u> - Install the distributor suppressor in the center socket of distributor head, as shown in Figure 8, by removing the high tension lead which runs from the distributor head to the coil and plugging the split end of

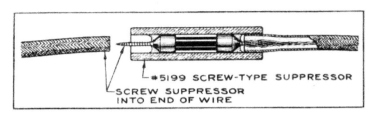

#5199 SCREW-TYPE SUPPRESSOR
SCREW SUPPRESSOR INTO END OF WIRE

Fig. 7

the suppressor into the distributor head, making sure of a good contact. Insert the high tension lead in the other end of the suppressor. If the car has a cap type distributor, the suppressor may be plugged in the coil or the screw type suppressor may be used, see Figure 7. In cars having two coils, a suppressor in each coil or high tension lead is necessary. Always install the suppressor as close to the distributor as possible.

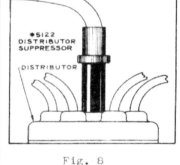

#5122 DISTRIBUTOR SUPPRESSOR

DISTRIBUTOR

Fig. 8

<u>GENERATOR CONDENSER</u> - Fasten the lead of one of the .5 microfarad condensers to the generator side of the cut-out relay on the car generator and clamp the condenser to the frame of the generator the screw holding the cut-out may be used for this purpose. Be sure that the condenser is securely fastened and a good ground connection made. In most cars, this condenser can be installed and connected as illustrated in Figure 9.

<u>AMMETER CONDENSER</u> - Fasten the lead of the other .5 microfarad condenser to the storage battery side of the ammeter. This usually is the terminal that has only one wire connected to it. Secure condenser to instrument panel (if it is metal) or to some metal part being sure of a good ground connection. A typical installation of this condenser is shown in Figure 10. Sometimes this condenser is more effective when attached to the dome light, stop light or horn wires. This latter is usually necessary when the car is equipped with a roof antenna. This may be tried while the motor is running and the effect on the interference noted; the condenser being connected to most effective point. It may be necessary in extreme cases to connect a condenser to more than one of these points in order to obtain reception from interference.

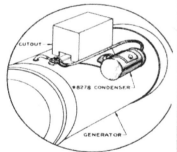

CUTOUT
#8278 CONDENSER
GENERATOR

Fig. 9

The above procedure will effectively suppress motor interference in practically all installations. However, if this does not hold true, it may be necessary to apply one or more of the following methods before complete suppression is obtained.

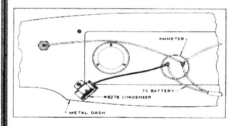

AMMETER
TO BATTERY
#8278 CONDENSER
METAL DASH

Fig. 10

First, determine whether the radiation is picked up by the antenna or by the receiver itself. This can be done by grounding the antenna lead as it leaves the receiver. If the motor interference stops, one may be sure that it is being picked up by the antenna. If it continues it is quite certain that part of the noise is being picked up by the receiver itself. If this is the case, make sure that all ground connections are clean and tight. If the instructions for installing the receiver have been carefully followed and all wires of the radio set have been kept out of the motor compartment, there should be no receiver pick-up. In the event of antenna pick-up of motor interference, the following suggestions are made to eliminate

GRIGSBY - GRUNOW CO.

it. These suggestions should be followed in the order in which they are given and the motor started and tested for interference after each step.

Peen the rotor.   It may be necessary to reduce the gap between the rotor arm and contacts of the distributor head.   Extreme care should be used in this operation to prevent harming the distributor. Peen the rotor by placing it on a flat steel block and hammering the end with a small machinist's hammer. Repeat this operation until there is just sufficient clearance - about .004". The rotor must not be allowed to touch the contacts. If there is evidence of the rotor touching the contacts, file off about .001" and recheck.   Building up the rotor arm with solder is not recommended as the solder is very soon burned away. In some cases, where the rotor is badly worn, it may be best to substitute a new one.

If the motor interference still continues, it may be well to determine the source. This can be done by removing the high tension lead from the coil to the distributor, turning on the ignition switch and cranking the car by hand. If a clicking is heard in the speaker, you may be sure that part of the trouble comes from the breaker points in the distributor or low tension circuit. It will then be necessary to remove the primary lead which runs from the coil to breaker points on the distributor, and replace it with a No. 14 shielded low tension cable, being sure not to run close to the high tension leads. The shielding must be grounded in at least two places. All ground connections must be as short as possible. It may be necessary to remove the lead from the switch to the coil and replace with a No. 14 shielded low tension cable being sure to ground the shielding. Care must be used when shielding so as not to short the coil or switch. Never use a by-pass condenser on this part of the circuit because it will effect the operation of the motor.

When you have tested to determine the source of motor interference and no clicking was heard in the speaker, we may assume that the interference is coming from the high tension or secondary circuit which is possibly the worst source of motor interference. All wires which run parallel to or within the field of this part of the circuit act as carriers and should be moved whenever possible, or the high tension wire re-routed. Sometimes the car manufacturer utilizes the high tension manifold to hold various wires and just removing them from the manifold will be sufficient. Be careful to keep the high tension lead as far as possible from the receiver. If after moving the wires, the interference continues, the high tension lead should be shielded. Care should be used when shielding the high tension lead to prevent the current from leaking through to ground. To prevent this first cover the high tension lead with loom, then run this shielding over the loom. The shielding must be grounded in at least two places (to the coil and motor block or high tension manifold). When the coil is under the cowl or bulkhead, the high tension lead should run as direct as possible to the motor compartment. This will sometimes necessitate drilling a new hole about one-half inch in diameter in the firewall or dash.

Due to the electro-magnetic field surrounding the ignition coil, it may be necessary, when the coil is under the cowl or bulkhead, to move it into the motor compartment. Mount it on the motor block as close to the distributor as possible and be sure that a good ground connection is maintained. If it is found necessary to mount the coil over the motor, care should be taken that it is so mounted as to stay sufficiently cool. New primary wires will be required and shielded No. 14 low tension ignition cable should be used. Caution: Do not run these wires close to the high tension lead, but ground them well.   ONLY MOVE THIS COIL AS A LAST RESORT.

In a number of cases, the establishing of a good electrical contact between the motor block, firewall and frame of the car will eliminate much of the interference. In assembling automobiles, oftentimes paint or other substances will prevent a good ground connection from being made between the various metal parts of the car which form the ground circuit. These poor connections will have no apparent effect on the operation of the car. However, when a radio receiver is installed, it is especially desirable to maintain all the metal parts of the car at the same ground potential. This is accomplished by connecting together with short pieces of shielding the motor block, frame, and firewall and sometimes the body of the car. Bonding may be particularly necessary on those cars having the motor mounted on rubber blocks. When bonding the motor to the firewall, use one inch shielding and make the bond long enough to allow for vibration of the motor.

GRIGSBY - GRUNOW CO.

Each and every wire, rod or pipe that runs from the motor compartment through the firewall into the body of the car acts as an antenna to radiate interference and should not be overlooked. To stop them from radiating, solder a heavy flexible copper conductor to them close to the firewall, allowing room for any movement of the rods, and then ground each of these to the firewall. If they are rusty, scrape them clean where contact is made. The wire conduit that runs to the base of the distributor on some cars should be grounded the same as other wires or rods.

In some instances, the noise being heard in the receiver will be caused by loose wires at the headlights, horn or horn button, tail light, stop light, stop light or dome lights. All connections to these items should be checked to see that the contact surfaces are clean and all wire connections are tight. Sometimes the connecting of a, Part #8278, .5 microfarad condenser to the hot battery lead feeding one or more of these accessories will have a decided effect on the ignition noise. This is especially true of the dome light wires when a roof type antenna is being used.

Any metal parts about the set making imperfect or intermittent contact with each other will cause noise in the speaker when the car is subjected to a jolt, whether there is any measurable potential difference between these parts or not. This interference is due to the instantaneous change in resistance of the receiver to ground that occurs when another ground conductor touches or is disconnected from the receiver.

To guard against the possibility of such noises, choke wires, speedometer cables, copper tubes, battery cables or the like should not be allowed to rub on the radio container. Also, always make sure that the chassis mounting bolts and the bolts that hold the brackets are securely tightened so that there will be no possibility of the contact resistance to ground changing.

If the foregoing information is insufficient to give complete suppression of motor interference, write in detail to the Majestic distributor in your territory or to the Grigsby-Grunow Company.

GRIGSBY - GRUNOW CO.

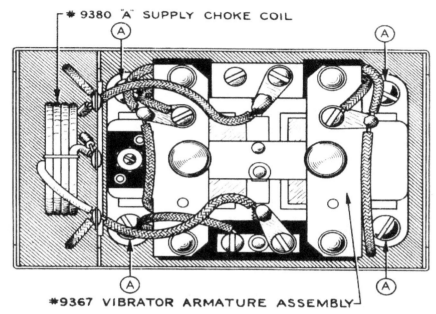

#9380 "A" SUPPLY CHOKE COIL

#9367 VIBRATOR ARMATURE ASSEMBLY

Fig. 11

DURO-MUTE POWER UNIT

The Duro-Mute Power Unit of the MAJESTIC Model 116 Auto Receiver is completely housed in the large metal container located at the extreme right of the receiver (see Figure 12).

Do not tamper with this unit unless it has proven defective by causing a gradual decrease in plate voltages and power output.

Should it, at any time, become necessary to inspect or replace the vibrator armature assembly of this unit, the procedure outlined below should be followed:

If the receiver is installed in the automobile, remove it from the firewall by loosening the clamping screws and sliding it off the supporting brackets.

Take off the top and bottom covers of the chassis container.

Unsolder the red, yellow, blue and black leads from the speaker output transformer (see Figure 12).

Remove the flexible drive cable from the gang condenser drive pulley, being careful not to cause any sharp bends or kinks in the cable.

After removing the five screws from the ends of the receiver, lift the container and speaker from the chassis, being careful not to place undue strain on the antenna lead wire.

Unscrew the four screws which hold the cover of the Duro-Mute Power Unit in place. The cover is easily removed by rocking slightly and lifting upward.

The entire vibrator armature assembly is now accessible for inspection or replacement.

WARNING!
Do not file the contacts or tamper with any of the adjustments on the vibrator armature

GRIGSBY - GRUNOW CO.

assembly. This unit has been carefully adjusted at the factory for utmost efficiency and any changes will seriously affect its operation.

The guarantee on the receiver will become void if the above warning is not followed.

If the vibrator armature assembly is known to be defective, remove it by disconnecting the necessary wires and unscrewing the four large screws marked "A" in Figure 11.

Replace with a new part #9367 vibrator armature assembly.

If there was a spacing washer under each of the screws at "A", they should not be used when the vibrator armature assembly is replaced with a new one.

Replace the Duro-Mute Power Unit cover, being certain that is fits snugly and properly supports the filter choke clamp.

Reassemble the outer container and speaker to the chassis and replace the bottom cover. Solder the speaker leads as indicated in Figure 12.

Assemble the flexible drive cables to the drive pulley so that with the tuning dial rotated to zero, the condenser gang will be completely unmeshed.

Turn on the receiver and test for proper operation over the entire tuning range, also noting that the drive cable operates smoothly and correctly.

Replace cover and assemble receiver to firewall.

CAUTION! Be sure to tighten all nuts and screws securely.

VOLTAGE CHART FOR MODEL 116 AUTO RECEIVER

TUBE	PURPOSE IN CIRCUIT	PLATE VOLTAGE	SCREEN VOLTAGE	CATHODE VOLTAGE	SUPPRESSOR VOLTAGE	GRID VOLTS
G57A-S	1st Detector Oscillator	110	110	15	0	1.4
G58A-S	1st I.F. Amplifier	180	90	3.5	3.5	...
G58A-S	2nd I.F. Amplifier	180	90	3.5	3.5	...
G75	2nd Detector and 1st Audio Amplifier	135	...	2.25	...	...
G89	Power Amplifier	170	180	0	0	...
G6-Z5	Rectifier	...	...	180	...	...

NOTE: All measurements made from designated points to ground with a 1000 ohm per volt, 300 volt range, D.C. voltmeter, the receiver connected storage battery delivering 6.0 volts at the battery terminals under load, the condenser gang fully meshed, and no signal supplied to the input of the receiver.

The tubes should be previously tested to assure that they are in good condition.

GRIGSBY - GRUNOW CO.

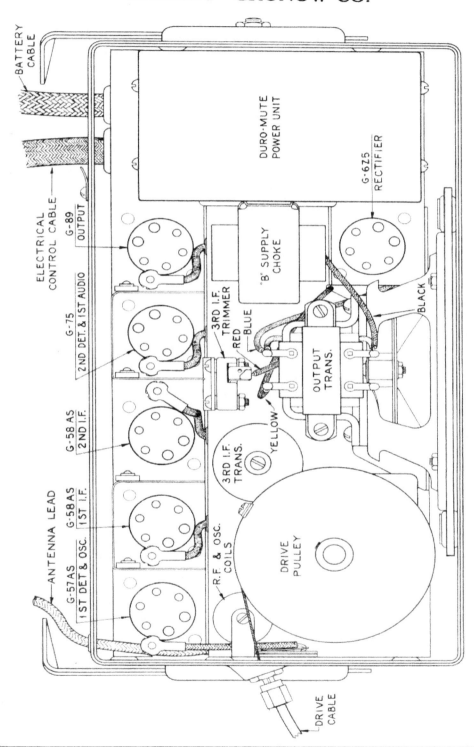

BATTERY CABLE

ELECTRICAL CONTROL CABLE

ANTENNA LEAD

G-57AS
1ST DET & OSC.

G-58AS
1ST I.F.

G-58 AS
2ND I.F.

G-75
2ND DET. & 1ST AUDIO

G-89
OUTPUT

DURO-MUTE POWER UNIT

G-625 RECTIFIER

"B" SUPPLY CHOKE

3RD I.F. TRIMMER

RED

BLUE

BLACK

OUTPUT TRANS.

3RD I.F. TRANS.

YELLOW

R.F. & OSC. COILS

DRIVE PULLEY

DRIVE CABLE

GRIGSBY - GRUNOW CO.

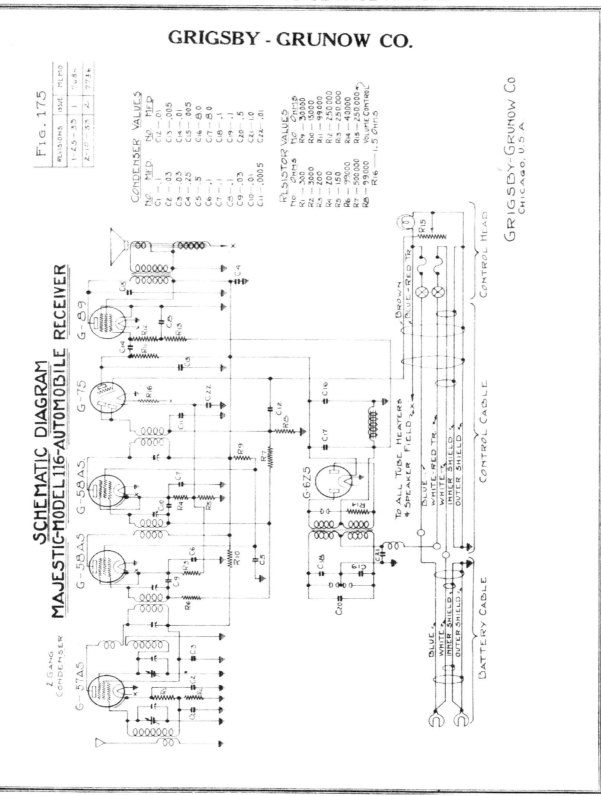

NATIONAL COMPANY, INC.

NATIONAL AUTOMOBILE RECEIVER

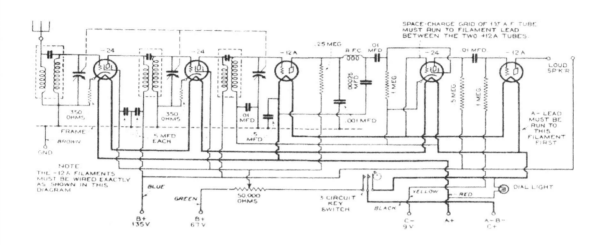

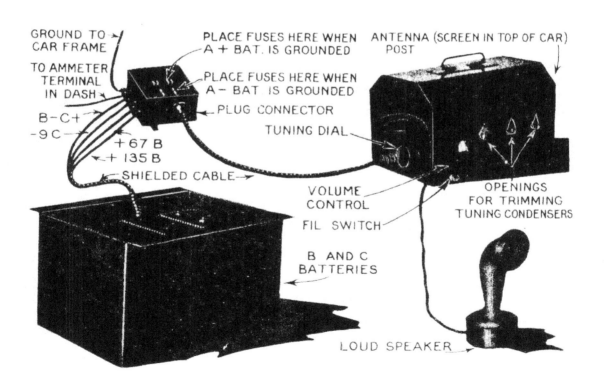

PHILCO RADIO & TELEVISION CORP.

TRANSITONE INSTALLATION and SERVICE BULLETIN FOR MODELS 7, 8 and 12

Standard Installation Procedure

Top Construction and Factory Antenna List

Car	Top Construction		Antenna		Lead-in Location
	Wood Slat	Poultry Screen	Wire	Poultry Screen	
*Auburn	V	....	.....	....	....
Buick............	V	....	...	....	....
Cadillac.........	V	....	V	....	Front rt. post
Chevrolet.......	V	....	.....	....	....
*Chrysler	Imp. 80	All others	Imp. 80	All others	Front rt. post
*Cord	....	....	.....	....	....
*DeSoto.........	....	V	.....	V	Front rt. post
*Dodge	....	V	......	V	Front rt. post
*Duesenberg ..	....	....	Special Bodies		....
Durant..........	....	V	.....	....	....
*Essex	Fabric	....	.....	....	.....
Ford A	....	V	.....	....	....
Ford 1932	....	V	......	V	No lead-in
*Franklin	....	V	.....	V	Front rt. post
Graham.........	V	....	.....	....	....
*Hudson	Fabric	Club Sdn.	.....	....	....
Hupmobile.....	....	V	.....	....	....
Jordan..........	....	V	.....	....	....
LaSalle.........	V	....	V	....	Front rt. post
*Lincoln	....	V	Tops are Cleared		....
Marmon.........	....	V	.....	V	Front rt. post
*Nash	....	....	V	....	Front rt. post
Oldsmobile......	....	....	.....	....	....
Packard.........	....	V	.....	....	....
Peerless........	....	V	.....	V	Front left post
*Pierce Arrow ..	....	V	.....	V	Front rt. post
*Plymouth	....	V	.....	V	Front rt. post
Pontiac.........	....	....	.....	....	....
*Reo	....	V	.....	V	Front rt. post
Rolls Royce	....	....	.....	....	....
*Studebaker	....	V	.....	V	Front rt. post
Stutz...........	....	....	.....	....	....
Willys-Knight...	....	....	.....	....	....

* These cars can be ordered from the car factory equipped with Philco Transitone.

By maintaining clearance between the poultry wire and the metal quarters of the body during the construction of the car, the car manufacturers have been able to build in a good car antenna. A few of the car factories install a wire antenna in the roof.

Installing an Antenna in Cars with Slat Top Construction—The headlining should be lowered from front to back so that a copper screen antenna can be installed in the roof.

1. Use a good grade of copper screen. No. 14 or No. 16 mesh, 36-inches wide is satisfactory and can be used in practically all installations.

2. Maintain three inches clearance between the screen and the car body and all metal work in the top. Cut out a section of the screen to get this clearance around the dome light.

3. The wiring in the top to the dome light and switch must be run along the side of the top frame, then along the top edge of the side of a bow to the dome light fixture.

4. An 18-gauge stranded copper, rubber and cotton covered antenna lead-in should be soldered to a front corner of the antenna screen. If the Receiver is to be located on the right side of the car, solder the lead-in to the right front corner of the antenna; if the Receiver is to be located on the left side, the antenna lead-in should be soldered to the left front corner. It is a good plan to solder or bond the whole front edge of the antenna screen.

5. The copper screen must be tacked securely so that it cannot come loose.

6. The headlining and all trim must be carefully replaced.

Tack the screen to the farthest bow in the rear that will give three inches clearance from the rear metal apron. With the edge of the screen lined up with the bottom front edge of the bow, the screen is tacked against the face of the bow, close to the top. It is necessary to tack the screen in this manner, so that the listing strip used to support the headlining can be tacked to the face of the bow.

On bows on which the listing strip is not tacked, it will be quite all right to tack the screen along the bottom of the bow. Tack the screen to each bow from the back to the front of the screen. Do not come closer than three inches to the metal aprons along the sides and the metal frame above the windshield.

The lead-in should be concealed behind the windshield moulding, or if the front corner post is hollow, it can be run down the inside of the post. In a few cases, it may be necessary to bring the lead-in down through the wind hose along the side of the corner post.

After the antenna and lead-in have been installed, test the antenna for grounds. Use a high resistance

PHILCO RADIO & TELEVISION CORP.

TRANSITONE INSTALLATION and SERVICE BULLETIN FOR MODELS 7, 8 and 12

volt-meter and a 45-volt battery, testing between the antenna lead-in and the body of the car. Do not hold the test connections to the antenna and the car body with your fingers,—as the leakage across your body will cause a high reading on the meter.

Having made certain that the antenna system is clear of grounds and leaks, proceed with replacing the headlining and trim.

Installing an Antenna in Cars with Poultry Wire Reenforcement—The poultry wire when cleared of grounds may be used as an antenna. This may be done in either of two ways. The top deck may be removed and the netting cleared where the edges ground on the car body. The more practical way is to drop the headlining the entire length of the car and clear from beneath.

A strip three inches wide is cut from the poultry wire reenforcement around the four sides. The poultry screen is then laced securely in place using double strands of number six waxed linen cord. Use short lengths of cord and fasten securely. The poultry wire must be held taut so the top will not sag. Care must be taken to keep the sharp ends of the screen bent back so they will not puncture the padding and the top deck material and will not extend through the headlining.

On standard installations, the antenna lead-in must be soldered across the front end of the screen and brought down the front right corner post. In cases where the post is solid, the lead-in may sometimes be brought down inside the windshield moulding or down the hollow rubber wind hose which is used in many cars.

Rearrange the dome light wiring so that there is a minimum coupling between the wires and the poultry wire antenna. Test the installation for grounds, using a 45-volt "B" battery and a high resistance voltmeter. Replace the headlining and trim carefully.

Installing an Antenna in Cars with Fabric Top Construction—In a few cars, the top padding is supported by muslin strips stretched over wood bows. An antenna can be easily installed in these cars in much the same manner used in cars with the slat top construction. Instead of tacking the screen under the bows however, the screen can be placed over the bows and tacked only at the rear and the front. Otherwise the procedure is the same.

Installing an Antenna in Cars with Metal Braces—In case there are metal diagonal braces in the top, the braces must be freed of grounds or the efficiency of the antenna will be greatly impaired.

Usually the rear ends of the braces are fastened to the wood top frame while the front ends are fastened by means of brackets to the front corner posts.

Drop the headlining and work from the inside of the car. Release the front end of the braces. Ream out the hole in the bracket and use fibre washers and sleeve bushings to insulate the cross brace bolts from the brackets.

Usually the dome light is connected to one of the braces. Disconnect the lead from the brace and run a new ground to the car body.

When both braces have been insulated, the antenna can be installed in the standard manner.

Open and Convertible Model Cars—The tops of the open and convertible models are designed to fold back. Since the antenna cannot in any manner interfere with this, a wire antenna is the only practical one.

Remove the top material and lay it back, leaving the side flaps in place. Secure a piece of top fabric, matching that removed, and fasten it properly in place over the cross ribs and over the side flaps.

Cut a piece of drill cloth or muslin approximately three inches smaller than the width of the top and about the length of it. Punch holes in the drill cloth through which the antenna wire is to be woven. The holes should be in rows, three inches apart, parallel to the cross ribs. Space the holes about ten inches apart in each row. In case metal bows are used, be sure to space the wires three inches from each bow.

Use 18-gauge stranded rubber covered wire and weave it back and forth through the holes in the cloth. When completed, the cloth is fastened to the front and rear bows only.

The antenna lead-in must be brought down in the rear so the top may be lowered easily.

The top material and all trim must be carefully replaced. While it is hardly probable that the antenna is grounded, check it with a voltmeter to make sure.

Receiver Installation—Install the Receiver on the inside of the dash, high and as far to the right as possible. Two sets of clinch-on nuts are pro-

PHILCO RADIO & TELEVISION CORP.

vided, one set on the back, the other on the left end, so that the Receiver may be mounted on the dash in either position. The end mounting will be found very convenient when a car is equipped with a hot water heater and not much room is available for the Receiver.

Using a template, mark the location of the bolt holes. Be sure to allow sufficient clearance for the Receiver. Center-punch and drill three small holes from the inside of the dash. Then drill again, using a ⅜-inch drill. This can usually be done from the engine side of the dash.

In case there is a vacuum tank or other apparatus near where the holes are to be drilled, remove the apparatus to avoid damaging it. Smooth off any burrs or rough edges on the holes. The paint on the dash around the holes should be scraped so that there will be good contact between the Receiver and the dash.

The Receiver which is being installed should be given a quick operating check as a precautionary measure. While the Receiver is still on the bench, remove the front cover plate. Place the plate and the screws to one side. Remove the corks from the set of clinch-on nuts which are to be used and then install the Receiver mounting studs in the Receiver. Place a 5/16-inch shake-proof lock washer on the short end of the stud and screw the bolt into the Receiver. The bolts should be fastened securely.

Install the Receiver on the dash, placing the large flat washers on the inside against the padding and the small washers against the metal side of the dash. The Receiver must be bolted securely to the dash.

In the Model "A" Ford, due to the location of the gas tank, it is necessary to mount the Receiver on the left side of the dash in the engine compartment.

Speaker—The speaker should be mounted on the inside of the dash over the steering column or toward the center. It should be placed high enough so that it will not interfere with the operation of any of the pedals or controls.

Using a template, mark the location for the two bolt holes. Be sure to allow sufficient clearance for the speaker housing. Center-punch and drill two small holes from the inside of the dash. Then drill again, using a ⁷⁄₁₆-inch drill. Install the speaker and bolt it securely to the dash. The tone control should be on the right.

In the Model "A" Ford, the Speaker must be mounted on brackets against the right kick pad. The brackets should be made up locally.

Bus and Boat Installations—The Model 12 is designed to operate from a 12-volt battery and is intended primarily for buses and motor boats.

In buses the Receiver and speaker will be installed in the most suitable location. In most installations, it will be possible to place the Receiver, speaker and Dynamotor in a metal container mounted on the baggage rack directly in back of the driver's seat.

In boats, the Receiver and dynamotor must be placed in the most convenient location available. The speaker can be bolted to a locker door or a wood partition in one of the cabins. The control unit usually must be located on a partition also.

Control Unit—The control unit has been designed so that it may be mounted in either of two positions on the steering column. The unit is compact, simple to install, easy to operate and has an artistic and well balanced appearance. It is shipped from the factory with two flexible shafts for the volume control and the tuning control coupled to it, although they may be removed very easily if it is ever necessary.

There is a mounting bracket which must be fastened to the steering column by means of a metal strap. This bracket should be installed in a horizontal position on the right side of the steering column, or in a vertical position above the steering column. Bend the metal strap around the steering column without using the felt pad. The round nut should be on the inside against the column. The strap should be fitted closely around the column and lapped over the end.

There are four small holes in the end of the strap. Cut off the excess strap about ⅜-inch beyond the hole that is to be used. Ream out the hole to ¼-inch.

Place the strap in position around the column again, this time placing the felt pad between the strap and the column. Fit the metal bracket against the column in the position desired, with the planed surface up. The fastening screw extends through the bracket and the hole in the strap and engages the round nut. Tightening the screw draws up the metal strap so that the bracket is clamped securely in place. Fasten the control to the bracket by means of the fillister head screw on the back of the housing.

PHILCO RADIO & TELEVISION CORP.

TRANSITONE INSTALLATION and SERVICE BULLETIN FOR MODELS 7, 8 and 12

The volume control and switch knob is on the left and is connected to the left hand flexible shaft. The tuning control knob is on the right and is coupled to the flexible shaft on the right. The black wire from the rear of the housing is the pilot lamp lead which must be connected to the Fahnstock terminal on the upper front edge of the Receiver. Dress the two flexible shafts and the wire neatly along the steering column and then up under the cowl. The two shafts should be held in place along the column by clamping them to the bottom edge of the instrument panel.

The volume control shaft must be fitted in the sleeve on the left hand side of the upper front panel of the Receiver. The shaft should be pushed in until the tip is all the way in the coupling on the volume control shaft. Fasten the casing by tightening the set screw on the bottom of the sleeve. This is inside of the Receiver.

With the volume control and switch knob turned off (in a counter-clockwise direction) and locked and with the volume control in the same position, tighten the bottom set screw in the coupling. Then rotate the shaft in a clockwise direction and tighten the other two set screws.

The tuning control shaft should be fastened in a similar manner. After dressing the shaft and fastening it in place, fit the shaft in the remaining sleeve in the upper front panel, and fasten the casing in place. The dial should be set at 55 and the condenser plates should be fully meshed. Tighten the bottom set screw, turn the shaft clockwise and then tighten the other two set screws.

After the flexible shaft is connected, a finer adjustment of the tuning condenser can be made so that the dial is properly lined up. This is done by tuning the Receiver to a station whose frequency is known. Check the scale to see how far off the dial setting is. If it needs changing, remove the face plate from the control housing. This is held on by two screws, one at the top and one at the bottom, which can be reached from the back of the housing.

Hold the tuning control to keep it from turning and lift the toothed edge of the scale over the teeth of the drive assembly and turn the dial to the proper setting. Then allow the dial to drop back in place so that the teeth on the dial mesh with the teeth on the drive assembly.

After the steering column control has been installed and the flexible shafts connected, replace the **front** cover plate on the Receiver.

Dynamotor—The Model EA Dynamotor is supplied as standard equipment with all Model **8** Receivers and the Model EC with the Model **12** Receivers. The Model **7** will be furnished with the Model EA Dynamotor in place of batteries when specified, or the Model EA can be ordered as a replacement unit for the Model **3** and Model **7** Receivers sold previously with batteries. The Model EA is for operation on 6 volt battery systems; the Model EC on 12 volt battery systems.

The dynamotor housing or box can be conveniently located in the floor of the car. Simply cut a hole $6\frac{1}{8}$ by $8\frac{7}{16}$ inches in the floor and drop the box in place from the top. Fasten the flange to the floor by means of screws or bolts.

It will be necessary to drill a hole in the end of the box for the battery cable. The tapered rubber bushing must be used over the hole to make it water-proof.

Model EA Terminal Arrangement

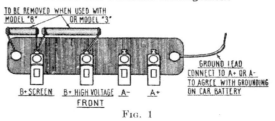

FIG. 1

When the Model EA is used with the Model **7** Receiver the blue lead must be connected to the "B+" High Voltage terminal and the green lead to the "B+" Screen terminal. The black-white lead must be connected to the relay switch which controls the operation of the dynamotor.

Looking at the top of the relay with the mounting bracket nearest the observer, the terminal on the side opposite the bracket must be connected to the car battery through a 15 ampere fuse. The terminal on the right must be connected to the "A" terminal on the dynamotor that corresponds with the live (non-grounded) side of the car battery. The remaining terminal on the left must be connected to the black-white lead of the battery cable. The relay should be mounted on the frame of the car near the car battery. The ground connection on the dynamotor and the shield on the cable must be connected to the other "A" terminal.

The dynamotor box must be grounded to the frame of the car by means of a heavy copper braid.

PHILCO RADIO & TELEVISION CORP.

TRANSITONE INSTALLATION and SERVICE BULLETIN FOR MODELS 7, 8 and 12

When used with the Model 8 Receiver, remove the two small fixed resistors at the left end of the terminal panel.

Connect the white-black lead to the "A" terminal on the dynamotor that corresponds with the live (non-grounded) side of the car battery. The ground lead on the dynamotor must be connected to the remaining "A" terminal. The cable shield must also be connected to this terminal.

Connect the blue lead to the "B+" High Voltage terminal. The dynamotor box must be grounded securely to the frame of the car by means of a heavy copper braid.

When the Model EA is used with the Model 3 Philco Transitone Receiver, remove the two resistors at the left end of the panel. The ground lead from the filter condenser must be removed from the ground terminal and must be spliced out and connected to the B+ Screen terminal.

The "B—" lead, the black lead which is grounded at the rear end of the dynamotor, must be removed from ground and must be spliced out and connected to the B+ Screen terminal also. This terminal now becomes "B—". Connect the blue-white lead to B+ High Voltage terminal and the green-white to B+ Screen terminal.

The relay switch must be used to control the dynamotor. With the relay in the same position as described above, the middle terminal must be connected to the car battery through a 15 amp. fuse. The terminal on the right must be connected to the "A" terminal on the dynamotor that corresponds with the live (non-grounded) side of the car battery. The remaining terminal on the left must be connected to the black-white lead of the battery cable. The relay should be mounted on the frame of the car near the battery. The ground connection on the dynamotor and the shield on the cable must be connected to the other "A" terminal.

The dynamotor box must be grounded to the frame of the car by means of a heavy copper braid.

The Model EC Philco Transitone dynamotor must be used only on a 12 volt battery system.

Connect the white-black lead to the "A" terminal on the dynamotor that corresponds with the live (non-grounded) side of the car battery. The ground lead on the dynamotor must be connected to the remaining "A" terminal. The cable shield must also be connected to this terminal.

Connect the blue lead to the "B+" High Voltage terminal. The dynamotor box must be grounded

securely to the frame of the car by means of a heavy copper braid.

Battery Box—The battery box is designed so that it can be installed in the floor of the car or suspended from it. In either case, check the location carefully so that there is sufficient riding clearance between the box and all the tie rods, braces, etc., on the chassis when the rear springs are depressed. Don't put the box too close to the exhaust pipe or the propellor shaft.

To install the box in the floor, after the proper location is found, cut a hole 10¼ x 8⅝ in the floor boards and drop the box in the hole so that it is supported by the flanges.

Drill two holes in the side of the box, a 9/16-inch hole for the small "A" cable bushing and a ¾-inch hole for the "A-B" cable bushing.

Fasten the box to the floor by means of four No. 8 ¾-inch wood screws. The holes for these screws are punched in the flange of the box but are covered by the cork gasket. These can be located and the cork punched out, before the box is installed.

When installing the box beneath the floor, after the proper location is found, drill two holes in the floor, 5/16-inch, for wood floors and ½-inch for metal floors. The flat bolt strap can be used as a template for drilling the holes.

After the holes are drilled, place this flat strap on the floor over the holes and push the long carriage bolts through from the top.

Holes must be drilled and the cable couplings installed, three Philco dry "B" batteries, P-302, placed in the box and connected,—and all cable connections made to the fuse mountings. Then the lid must be screwed down tightly, so that the entire box is water-tight.

Push the battery box up against the floor with the bolts extending through the square holes in the flange. Put the box support on next, with a bolt passing through the hole in the end. Run a nut up on both bolts and slip the slotted end of the support over the other bolt and nut. After tightening both nuts, put on a lock nut and a cotter pin in the end of each bolt.

The battery box is shipped from the factory with a cardboard liner inside it. This liner must be left in the box. After the cable couplings are installed, place three P-302 Philco dry "B" batteries in it and connect them in series. The battery cable should

PHILCO RADIO & TELEVISION CORP.

TRANSITONE INSTALLATION and SERVICE BULLETIN FOR MODELS 7, 8 and 12

then be cut off at the proper length and connections made to the fuses and batteries.

The blue-white lead must be connected to B+135 volts and the green-white lead to B+67½ volts. The black-white wire is the "A" lead which must be connected to the ten-ampere fuse and from there, a single lead which is supplied, must be run to the car battery. "B"— of the batteries must be connected through the one-ampere fuse to the battery side of the ten-ampere "A" fuse.

The bakelite fuse mountings should be screwed to the wood hold-down which is placed over the batteries. Before the lid is put on, the flaps of the liner should be folded over the hold-down.

Three heavy duty Philco batteries, P-308, can be used in place of the standard size batteries and will give relatively longer life. Use the large box and lid (04585) and place one battery in upright and the other two on their sides with the terminals in the center. Pack the batteries to prevent bumping around.

Cable Connections—MODEL 7—The speaker and battery plugs must be connected in their respective sockets on the front panel and the ground tabs from the cable shields grounded under a screw head. The cable should be dressed and fastened in place.

The battery cable should be run down in back of the right kick pad and then through the floor and along the frame to the "B" battery box or dynamotor. It should be clamped in place securely.

Cable Connections—MODELS 8 and 12—The speaker and battery cables and the antenna lead are all formed in a one piece cable and are totally shielded. Connect the battery plug (female) to the socket on the Receiver. The speaker plug is on the end of the cable and must be connected to the socket on the side of the speaker housing. Ground the shield pigtails on these cables on the speaker and Receiver housings.

The antenna lead is the next leg of the cable. Connect this to the antenna lead-in as close to the corner post as possible. Solder and tape the splice and then ground the shield pigtail on the bottom edge of the instrument panel. For best results, the shield should extend up into the corner post.

The shielded black-white lead must be connected through the 15 amp. fuse to the live side of the storage battery and the shielding grounded to the frame of the car or to the grounded terminal of the battery.

The Model 8 is for operation on 6 volt battery systems,—the Model 12 on 12 volt battery systems.

The end of the cable must be run to the dynamotor and connected properly. The black lead, the cable shield, must be grounded to the terminal panel. Keep the cables out of the motor compartment.

Suppression—The standard spark plug resistors 4531 can be installed on the plugs in most cars. Likewise the standard distributor resistor 4546 can be used in the distributor head in most cases.

On cars such as the Buick, where the standard spark plug resistor cannot be used, the special screw type resistor 4851 should be used. In a few cars, it will be necessary to use it at the distributor head also. When using the latter resistor, be sure it is as near as possible to the end of the lead.

Standard suppression calls for the use of one resistor on each spark plug or in the plug end of each lead and one resistor at the distributor in the high tension coil to the distributor lead.

In the case of a two coil system, two resistors are necessary, one in each high tension coil lead at the distributor. When dual ignition is used, each spark plug must be equipped with a resistor.

There are numerous exceptions to the above. If the radio installations are carefully made, it will be possible in many Buick and Cadillac installations to do without the spark plug resistors, using only one resistor in the distributor head. On the new Ford V-8, no resistor is used in the distributor head.

In addition to the standard use of resistors, two 4522 Condensers are also required, one on the brush side of the generator cut-out, the other on the battery terminal of the ignition coil.

When installing an interference condenser, connect the lead to the apparatus terminal. The bracket of the condenser must be bolted to the engine or some other grounded metal part of the chassis.

The use of resistors on the spark plugs and distributor head, and of condensers on the coil and generator, is termed standard suppression and is required in most installations.

In a great many cases, when radio installations are made by the car factories, radio spark plugs are used. These are regular spark plugs with the resistor unit built in the plug and sealed. In addition to making a very neat installation on the motor, their performance is entirely satisfactory.

PHILCO RADIO & TELEVISION CORP.

Plugs of this character are invariably marked "radio." Don't install the standard resistors on plugs of this type.

Peening the Rotor Arm—Quite frequently it is necessary to peen the rotor in the distributor in order to reduce the gap between the rotor and the high tension contacts. The gap should be held to about .004 inches maximum, but care should be taken that the rotor does not brush any of the contacts.

Place the rotor on a flat steel block and hammer the end of the rotor carefully with a small machinist's hammer. Repeat this operation until there is just sufficient clearance between the rotor and the contacts. Using a file, dress the end of the rotor to its original shape. If a double end rotor is used, both ends should be treated alike, completing the operation first on one end, then the other.

Extreme caution should be used in this operation so that the distributor will not be damaged. Never pass an installation if the rotor brushes the contacts, as this affects the timing.

On the Ford V-8, the rotors cannot be removed easily, so instead of peening the rotors, build them up with solder.

Shielding—In the past, a great number of service men were prone to shield the high tension leads indiscriminately. This gave rise to numerous complaints on the car performance.

There is no need for shielding the high tension leads. The only possible exception to this is when the coil is mounted on the instrument panel. Relocate the high tension cable if necessary, so that it goes direct from the coil through the dash. Shield the lead by covering it with a standard length of shielded loom,—L-1387.

Shielding—Antenna Lead—All Receivers are now wired with a shielded antenna lead. This must be spliced to the antenna lead-in as close to the corner post as possible. Avoid all excess slack. It is advisable to continue the shielding up into the corner post for an inch or so. Ground the shielding at the corner post.

Additional Suppression—The intense high frequency field present under the hood is sometimes carried beyond the dash by pipe lines, rods and wires. To prevent this, some precautions are necessary.

In case of severe motor interference, isolate the high tension leads from the rest of the car wiring. Remove the low tension wires to the coil, horn wires or other cables from the high tension manifold.

Additional interference condensers may sometime be needed on fuse blocks, on the ammeter, or possibly on the dome light lead where it enters the front corner post. It is more important to by-pass the dome-light lead at the corner post than to connect the condenser to the ignition coil in some installations.

Always connect the "A" lead to the car battery. Unnecessary interference will be encountered if the "A" lead is connected elsewhere.

Occasionally it will be necessary to bond the dash to engine block. Use heavy copper braid for this, bolting the braid to both the dash and the engine block.

Use a smaller copper braid for bonding rods and pipe lines, fastening the braid to the dash with self tapping screws, and soldering the other end to the parts to be bonded. Keep all bonds as short as possible, but allow sufficient slack so as not to interfere with the operation of choke rods, etc.

Adjusting the R. F. Padding Condensers—In order to obtain the maximum results from the radio installation, the first and second R. F. padders should be adjusted after the installation is completed. This should not be attempted except by a competent service man.

It will be necessary to remove the front cover plate and to set up a good oscillator capable of generating a signal of approximately 1400 K. C. The Philco Oscillator, Model 095, can be used very satisfactorily for this adjustment as well as all other adjustments on the Receiver. Connect a six foot lead to the oscillator output terminal, simply dropping it over the back of the seat, and turn on the oscillator. Turn on the Receiver and tune to approximately 140 on the Receiver scale. Adjust the oscillator frequency to 1400 K. C. When using the Philco oscillator, set it for the 175 K. C. range and use the eighth harmonic. Turn on full volume on the Receiver and adjust the output of the oscillator until the signal is barely audible. Tune the Receiver sharply to the signal and then adjust the first R. F. padder. This is the one mounted to the extreme right on the condenser housing. Adjust this for maximum signal and then proceed with the second padder, the one in the

PHILCO RADIO & TELEVISION CORP.

TRANSITONE INSTALLATION and SERVICE BULLETIN FOR MODELS 7, 8 and 12

center. Use only the standard fibre padding wrench. Replace the front panel and the adjustment is completed.

Service—Philco Transitone products are designed and built to give the greatest owner satisfaction possible. Don't jeopardize the performance or the name of these products by poor service or the use of other than genuine parts.

Lack of knowledge of the product, incorrect procedure, careless workmanship, lack of courtesy in dealing with the customer or an inadequate stock of replacement parts, tubes, batteries and testing equipment will result in poor service and in actual loss in business.

Learn everything possible in connection with the Receiver and the correct installation procedure. Give the customers more attention than they expect. Don't tolerate rudeness on the part of the installation station employees. See that you have a complete stock of all Philco Transitone parts required for prompt service in case it is needed.

Don't overlook the replacement tube and battery business. There is a Philco tube or battery for every purpose. Always have enough on hand. There is also a growing market for suppression material. Every motor radio needs suppression material to complete the installation and make it satisfactory. Philco Transitone installation stations should get this business.

Installation stations should also find a good market for the dynamotor, to be used with other Receivers as well as the former Philco Transitone models. All this extra business is within the reach of any good installation station that renders proficient service.

Special Adjustments—In order to render proficient service, the installation station must be able to make the proper adjustments to the Receivers whenever they are needed. This is impossible without the use of a good service oscillator. The best and most economical oscillator for this work is the Philco Oscillator, Model 095. Complete information and instructions for its use can be had on request from your Philco Transitone distributor or from the service department at the factory.

The adjustments should be made as follows:

Intermediate Frequency or I. F. Stages—Remove the grid clip from the detector oscillator tube and connect the output of the oscillator to the control grid. The detector oscillator is the second tube from the right.

With the Receiver and oscillator turned "on", set the oscillator for 175 K. C. Adjust the oscillator attenuator so that the signal is barely audible with the Receiver volume control turned on full. If the oscillator is equipped with an output meter, connect the meter and adjust the attenuator so that a half scale reading is obtained.

Using a Philco 3164 fibre wrench, adjust the second I.F. condenser. This is numbered ⑱ on figs. 3 and 5 and ㉙ on figs. 4 and 6.

The correct adjustment is obtained when the strongest signal is heard in the speaker or the maximum reading is secured on the meter.

Next adjust the secondary and primary I.F. condensers. These are ⑲ and ⑬ respectively on figs. 3 and 5 and ㉒ and ⑲ on figs. 4 and 6.

Disconnect the oscillator and reconnect the clip to the control grid.

High Frequency Compensator—Connect the output of the oscillator to the antenna lead and the housing of the Receiver. With the Receiver turned on and the oscillator set for 175 K. C., tune the Receiver to 1400 K. C., the eighth harmonic of 175 K. C., and adjust the third padder on the tuning condenser for maximum signal. This is the one on the extreme left of the housing. The purpose of this adjustment is to line up the condenser so that 1400 K. C. is tuned in at 140 on the scale when the scale is set properly.

It may be necessary to adjust the first two compensators on the tuning condensers at 1400 K. C., in order to get a strong enough signal through.

R. F. Compensators—After the detector oscillator has been padded at 1400 K. C., adjust the first and second R. F. Condensers on tuning condenser at 1400 K. C.

Low Frequency Condenser—Now tune the Receiver to 700 K. C. and adjust the condenser ⑭ on figs. 3 and 5 and ⑰ on figs. 4 and 6. During this operation the tuning condenser must be shifted and the compensators must be adjusted to bring in the maximum signal.

After this has been done, check the adjustment of the high frequency condenser at 1400 K. C. again.

PHILCO RADIO & TELEVISION CORP.

TRANSITONE INSTALLATION and SERVICE BULLETIN FOR MODELS 7, 8 and 12

Model EA Dynamotor Wiring Diagram

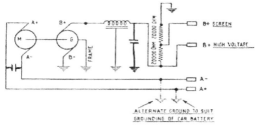

FIG. 2

Table 1—Resistor Data

Nos. on Figs. 3 and 5	Nos. on Figs. 4 and 6	Resistance (Ohms)	Color		
			Body	Tip	Dot
(46)	(10)	2.7		wire resistor	
(38)	(44)	7		" "	
...	(50)	30		" "	
(21)	...	225		" "	
...	(23) (7)	500		" "	
(38)	(42)	700		" "	
(7) (11)	(1) (15)	5,000	Green	Black	Red
...	(24) (26)	20,000	Red	Black	Orange
(31)	(37)	50,000	Green	Black	Orange
(3) (23) (24)	(3) (27) (28)	99,000	White	White	Orange
(20) (26)	(21) (31)	490,000	Yellow	White	Yellow

Table 2—Condenser Data

Nos. on Figs. 3 and 5	Nos. on Figs. 4 and 6	Capacity (Mfd.)	Color
(28)	(30)	.00025	Yellow
(10) (15)	(14) (18)	.0007	White and Yellow
(28) (34)	...	.001	Green and White
...	(33) (35)	.00125	Blue and Orange
(37)	(45)	.002	Blue
(33)	(39)	.01	Black Bakelite
(4) (18)	(5) (7) (50)	.05	Black Bakelite
(29)	(36)	.25	Metal Can
...	(11)	.25, .5	Metal Can
	See Note 1	.25, .25, .5	Metal Can
(16)	...	.25, .5, 20.0	Metal Can

PHILCO RADIO & TELEVISION CORP

TRANSITONE INSTALLATION and SERVICE BULLETIN FOR MODELS 7, 8 and 12

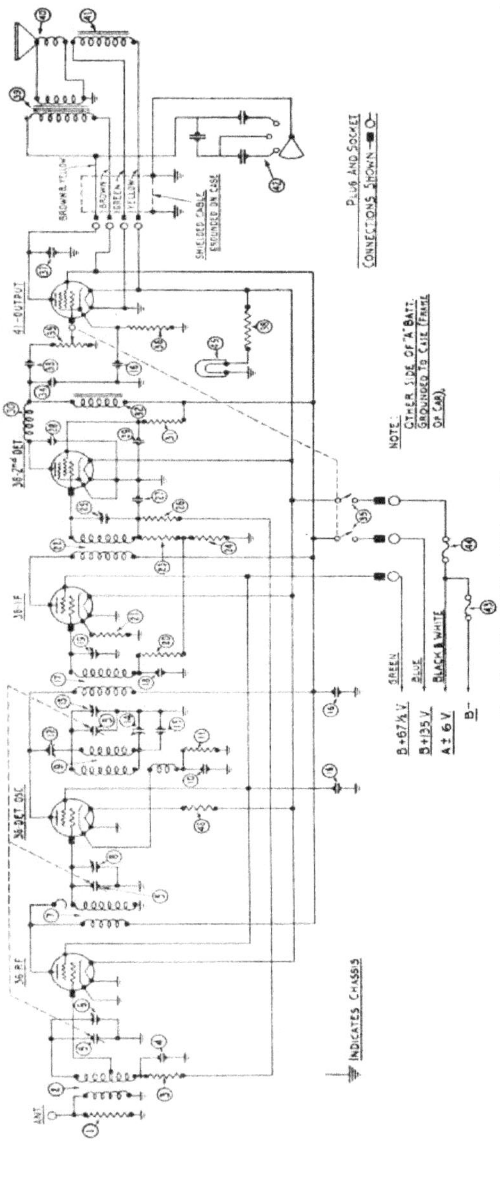

FIG. 3—Model 7—Wiring Diagram

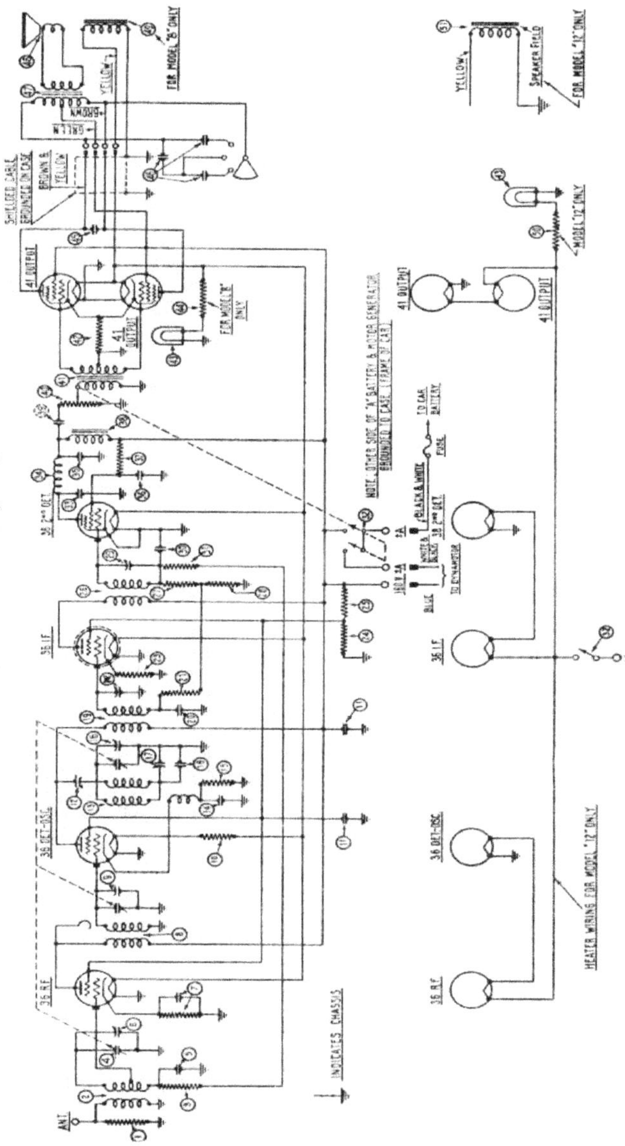

FIG. 4—Models 8 and 12—Wiring Diagram

PHILCO RADIO & TELEVISION CORP.

TRANSITONE INSTALLATION and SERVICE BULLETIN FOR MODELS 7, 8 and 12

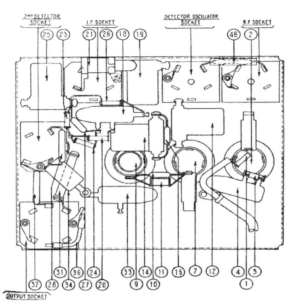

Fig. 5—**Model 7**—Chassis

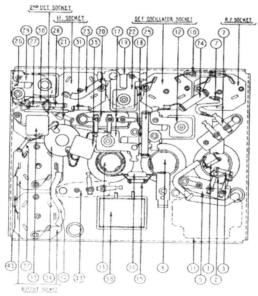

Fig. 6—**Models 8 and 12**—Chassis

PHILCO RADIO & TELEVISION CORP.

TRANSITONE INSTALLATION and SERVICE BULLETIN FOR MODELS 7, 8 and 12

Replacement Parts Models 7, 8 and 12

Models 8-12 Figs. Nos. 4-6	Model 7 Figs. Nos. 3-5	Description	Part No.
①	①	Resistor (5,000 Ohms)........	6096
②	②	Antenna Coil..............	04348
③	③	Resistor (99,000 Ohms).......	6099
④	⑤	Tuning Condenser Assembly...	04308
	④	Condenser (.05 Mfd.)........	3615-AG
⑤		Condenser (.05 Mfd.)........	3615-AN
⑥	⑥	Compensating Cond. (Part of Tuning Condenser)........	
⑦		Condenser (.05 Mfd.)........	3615-AE
⑦		Resistor (500 Ohms)........	6977
⑧	⑦	Detector Transformer........	04509
⑨	⑧	Compensating Cond. (Part of Tuning Condenser)........	
⑩	⑯	Resistor (2.7 Ohms).........	6511
⑪		Condenser (.25, .5 Mfd.).....	04959
		See Note 1 (.25, .25, .5 Mfd.)..	05622
	⑯	Condenser (.25, .5, 20. Mfd.).	04354
⑫	⑫	Compensating Condenser.....	04000-A
⑬	⑨	Oscillator Coil.............	04508
⑭	⑩	Condenser (.0007 Mfd.)......	4520
⑮	⑪	Resistor (5,000 Ohms)......	6096
⑯	⑬	Compensating Cond. (Part of Tuning Condenser)........	
⑰	⑭	Compensating Condenser....	04000-R
⑱	⑮	Condenser (.0007 Mfd.).....	5863
⑲	⑰	First I.F. Transformer.......	04352
⑳	⑱	Condenser (.05 Mfd.).......	3615-AK
㉑	⑳	Resistor (490,000 Ohms).....	6097
㉒	⑲	Compensating Condenser....	04000-D
㉓		Resistor (500 Ohms)........	9042
	㉑	Resistor (225 Ohms)........	6107
㉔		Resistor (20,000 Ohms)......	6650
㉕		Resistor (20,000 Ohms)......	6649
㉖	㉒	Second I.F. Transformer.....	04353
㉗	㉓	Resistor (99,000 Ohms)......	6099
㉘	㉔	Resistor (99,000 Ohms)......	6099
㉙	㉕	Compensating Condenser....	04000-A
㉚	㉗	Condenser (.00025 Mfd.).....	3082
㉛	㉖	Resistor (490,000 Ohms).....	6097
㉜		Switch (See Note 2)........	
㉝		Condenser (.00125 Mfd.).....	5886
	㉘	Condenser (.001 Mfd.)......	5215
㉞	㉚	R.F. Choke............	04342
㉟		Condenser (.00125 Mfd.).....	5886
	㉞	Condenser (.001 Mfd.)......	5215
㊱	㉙	Condenser (.25 Mfd.) See Note 3	04360
㊲		Resistor (50,000 Ohms)......	6098
	㉛	Resistor (50,000 Ohms)......	4237
㊳		Audio Choke............	6602
	㉜	Audio Choke............	5930
㊴	㉝	Condenser (.01 Mfd.)......	3903-Y
㊵		Volume Control (Note 2).....	7322
	㉟	Volume Control...........	6109
㊶		Input Transformer..........	6582
㊷	㊱	Resistor (700 Ohms)........	6443

Models 8-12 Figs. Nos. 4-6	Model 7 Figs. Nos. 3-5	Description	Part No.
㊸	㊺	Pilot Lamp................	4567
㊹	㊳	Resistor (7 Ohms)..........	5110
㊺	㊲	Condenser (.002 Mfd.).......	6583
㊻	㊸	Tone Control..............	05366
㊼		Output Transformer........	2565
	㊴	Output Transformer........	2598
㊽	㊵	Cone and Voice Coil........	02823
㊾	㊶	Field Coil Assembly (6V)....	02794
㊿		Resistor (30 Ohms)..........	7155
㊸		1 Amp. Fuse..............	4540
㊹		10 Amp. Fuse.............	5676
		15 Amp. Fuse.............	7227
51		Field Coil Assembly (12V)....	02688
		Battery Cable (Model 7).....	04416
		Battery Cable (Model 8-12)...	05419
		Plug (Model 7).............	4539
		Cap (Model 7).............	4885
		Plug (Model 8).............	7122
		Cap (Model 8).............	7123
		Fibre Wrench.............	3164
		Control Unit Assembly.......	04343
		Control Housing Cover......	6030
		Key (Interchangeable)......	6091
		Speaker Extension Cable....	02984
		Spark Plug Resistor.........	4531
		Distributor Head Resistor....	4546
		Special Resistor (Screw Type).	4581
		Interference Condenser......	4522
		Philco I. F. Oscillator...Model	095
		Type 36 Tube..............	5582
		Type 38 Tube..............	5584
		Type 41 Tube..............	6446
		Knobs..................	5166
		Receiver Housing...........	6058
		Speaker Housing...........	2710
		Dynamotor Complete—Model EA	05388
		Dynamotor Complete—Model EC	05424
		Dynamotor Only 6V........	6651
		Dynamotor Only 12V........	7165
		Dynamotor Filter Choke.....	6658
		Dynamotor Filter Condenser..	05386
		Dynamotor Housing.........	6655
		Large Battery Box (Complete).	04585
		Small Battery Box (Complete).	04581
		Receiver Studs.............	6122
		Shielded Loom............	L-1387

Description	Part No.
18″ Volume Control Shaft...............	6351
18″ Tuning Control Shaft...............	6352
32″ Volume Control Shaft...............	6128
32″ Tuning Control Shaft...............	6129
48″ Volume Control Shaft...............	6298
48″ Tuning Control Shaft...............	6299
120″ Volume Control Shaft...............	6355
120″ Tuning Control Shaft...............	6356

NOTE 1—In some Receivers, 04959 is replaced by 05622. ㊴ is omitted and a .25 Mfd. section of 05622 is used in its place.

NOTE 2—Switch ㉜ in fig. 4 is integral part of volume control ㊵, part No. 7322.

PHILCO RADIO & TELEVISION CORP

PHILCO TRANSITONE MODEL 5

The Philco Transitone Model 5 is the latest Philco development in automobile radio. It is a powerful and extremely compact superheterodyne having many of the features of the larger auto radio Receivers.

The Receiver, Speaker and the new Full Wave Philco Vibrator are all housed in a single shielded container designed for quick installation on the dash of the automobile. The arrangement is particularly adaptable for small cars and for cars already equipped with a heater. The full powered, electro-dynamic speaker is mounted in the bottom of the housing so as to afford excellent tone quality and volume without the necessity of using a speaker as a separate unit.

All the tubes used are the latest Philco high-efficiency tubes, designed especially for automobile radio. Several of these tubes each perform the functions which formerly required two and three tubes, thereby effecting a great tube economy, reducing the number of tubes necessary for satisfactory operation, and reducing the amount of current taken from the car battery to the very minimum.

Philco's system of automatic volume control is used to give that smooth, elastic control which counteracts fading while driving along, and prevents blasting of local stations.

The new Receiver is All Electric, operating entirely from the car storage battery. The new Full Wave Philco Vibrator is built in as an integral part of the Receiver.

This Receiver is destined to be one of the most popular models we have ever offered to the public and will meet with instant approval from everyone as soon as it can be seen and heard. Ease of installation will enable service stations to cut their costs and speed up installations. Customers will wait for their cars while the installation is being made, since the average installation will be made in only a fraction of the time formerly required.

THE NEW MODEL 6F

The Model 6 with the latest improvements becomes the Model 6F. The new features are entirely for your benefit.

Greater ease of installation and service is accomplished, cutting the installation time practically in half.

The Model 6F is now a two-piece unit instead of having the customary three pieces. The new improved full wave EF Vibrator is securely attached to the side of the Receiver housing. Drilling three holes for the Receiver and installing it, automatically takes care of the Vibrator installation.

Connections, too, are much more simple. The battery cable plugs into the Receiver receptacle, while a leg of the cable connects by a plug to the Vibrator unit. Just one lead to connect to the battery and the antenna lead. Install the Receiver and connect the cable.

If the Vibrator requires servicing, remove four screws from the lid and the Vibrator lifts right out of the housing. The only service it should ever require will be either a replacement tube or a replacement Vibrator unit. Don't ever attempt to adjust the Vibrator.

The circuit arrangement of the Vibrator is shown in the schematic, Fig. 3.

MODEL EG VIBRATOR

The Model EF Vibrator is a part of the Model 6F Receiver. Its counterpart for "B" battery replacement service is the Model EG Vibrator. Instead of being connected with a cable and plug, it is equipped with a terminal panel for easy installation.

When used as a replacement unit for "B" batteries, simply install in the old "B" battery box or in any place that is convenient and where the Vibrator will not be exposed to water and dirt. The installation is easy, but at the same time permanent.

Simplicity in construction insures freedom from trouble and efficient operation. Cut disc tungsten points eliminate any possibility of troubles from contacts. Full wave rectification with the 84 rectifier tube developed especially for this type of service is used to give a smooth flow of power. Complete filtering eliminates all hum.

Schematic and Base Arrangement of Philco Tubes used in Models 5, 6, 7, 8, 9, 12
Philco Transitone Receivers

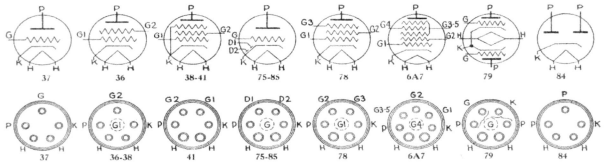

SYMBOLS—H = Heater, K = Cathode, G = Grid, P = Plate, D = Diode Plate.

PHILCO RADIO & TELEVISION CORP.

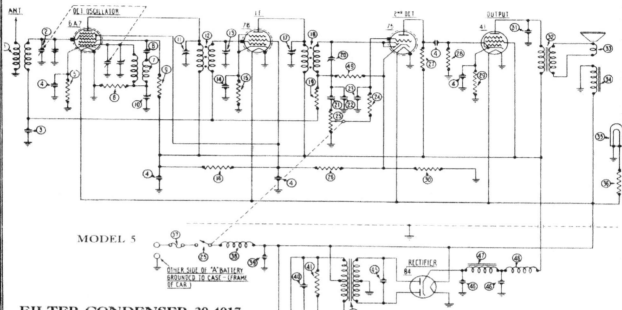

MODEL 5

FIG. 1

FILTER CONDENSER 30-4017

④ on Figs. 1 and 2

There are five sections in this filter condenser, all terminated with wire leads. The two green leads connect to the .1 mfd. section, which is used for coupling the plate output of the 75 tube to the grid of the 41 tube.

The remaining four sections are all grounded to the can on one side. The white leads connect to two .25 mfd. sections. The first section is connected to the cathode of the 6A7 tube. The second section is connected to the screen of the 78 tube.

The red lead from the .5 mfd. section is connected to the B+ side of all the plate circuits. A 20 mfd. section terminates in a black lead, which in turn is connected to the cathode of the 41 tube.

FILTER CONDENSER 30-4010

㊻ on Figs. 1 and 2

This condenser consists of two sections, a 4 mfd. section and an 8 mfd. section, both of them grounded on one side.

The 4 mfd. section terminates in a red lead, which is connected to the cathode of the 84 tube. The 8 mfd. section terminates in a green lead, which is connected between the two chokes in the rectifier filter circuit.

NOTE.—The first condensers (30-4017) were made up having five sections. The .1 mfd. section has now been removed from the can and this section replaced with a .006 mfd. condenser (Part No. 30-1001). This condenser is located in the chassis adjacent to the grid terminal of the 41 tube socket.

MODEL 5

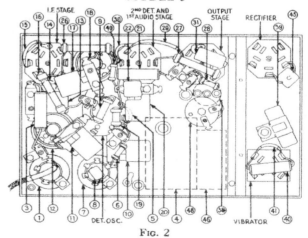

FIG. 2

PHILCO RADIO & TELEVISION CORP.

MODEL 5 PARTS LIST

No. on Fig. 1 and 2	Description	Part No.
①	Antenna Transformer	32–1084
②	Tuning Condenser	31–1019
③	Condenser (.05 mfd.)	30–4020
④	Filter Condenser (.25; .25; .5; 20 mfd.)	30–4017
⑤	Resistor (200 ohms)	7217
⑥	Resistor (1300 ohms)	8267
⑦	Oscillator Coil	32–1085
⑧	Condenser (.00025 mfd.)	3082
⑨	Resistor (15,000 ohms)	6208
⑩	Padder	04000–S
⑪	Padder	04000–J
⑫	First I. F. Transformer	32–1086
⑬	Padder	04000–Y
⑭	Condenser (.5 mfd.)	30–4018
⑮	Resistor (1,000 ohms)	33–3017
⑯	Resistor (10,000 ohms)	4412
⑰	Padder	04000–D
⑱	Second I. F. Transformer	32–1087
⑲	Resistor (1,000,000 ohms)	4409
⑳	Padder	04000–M
㉑	Condenser (.05 mfd.)	30–4020
㉒	Condenser (.00025 mfd.)	3082
㉓	Condenser (.0005 mfd.)	3910
㉔	Resistor (100,000 ohms)	6099
㉕	Volume Control and Switch	33–5009
㉖	Resistor (32,000 ohms)	3525
㉗	Resistor (250,000 ohms)	3768
㉘	Resistor (500,000 ohms)	6097
㉙	Resistor (700 ohms)	6443
㉚	Resistor (400 ohms)	33–3016
㉛	Condenser (.006 mfd.)	30–1002
㉜	Output Transformer	32–7005
㉝	Cone	36–3027
㉞	Field Coil	9013
㉟	Pilot Lamp	6608
㊱	Resistor (7 ohms)	7155
㊲	Fuse, 15 A.	7227

No. on Fig. 1 and 2	Description	Part No.
㊳	R. F. Choke (Low voltage)	32–1083
㊴	Condenser (.5 mfd.)	30–4015
㊵	Condenser (.05 mfd.)	30–4020
㊶	Resistor (200 ohms)	7217
㊷	Vibrator	38–5036
㊸	Resistor (200 ohms)	7217
㊹	Transformer	32–7030
㊺	Condenser (.006 mfd.)	30–1002
㊻	Condenser (4 mfd.; 8 mfd.)	30–4010
㊼	Filter Choke	32–7026
㊽	R. F. Choke (High voltage)	32–1078
㊾	Resistor (250,000 ohms)	4410
	Control Shaft (Tuning)	28–8006
	Control Shaft (Volume)	28–8007
	Tube Kit	34–3006
	75 Tube	8002
	78 Tube	8315
	41 Tube	6446
	84 Tube	34–2001
	6A7 Tube	34–2002
	Dial	27–5006
	Antenna Lead	L–1594
	Battery Cable (Bat. end)	38–5124
	Battery Cable (Rec. end)	38–5123
	Fuse Housing	28–1269
	Male Cap (Fuse)	28–1270
	Contact (Fuse)	27–7133
	Washer	27–7132
	Spring	28–8009
	Fuse Insulator	27–7131
	Antenna Male Cap	28–1270
	Contact (Antenna)	28–7133
	Spark Plug Resistors	4531
	Dist. Resistors	4546
	Screw Type	4851
	Interference Condenser (1 mfd.)	4522
	Interference Condenser (½ mfd.)	30–4007

MODEL EF FULL WAVE VIBRATOR
(Used With Model 6F Receiver)

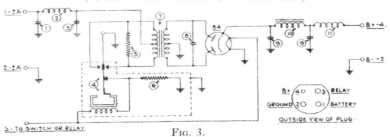

Fig. 3.

The Model EF takes the place of the EB dynamotor. The cable connection between the Vibrator and the Model 6F completes the installation of the Vibrator. Terminal 1 is connected directly to the main battery lead. Terminal 2 is the cable shield. Terminal 3 is connected to the Radio switch. Terminal 4 is the B+ high voltage lead and is connected directly to the plate circuits.

MODEL EF—PARTS LIST

No. on Fig. 3	Description	Part No.
①	Condenser (.5 mfd.)	30–4015
②	R. F. Choke (Low voltage)	32–1083
③	Condenser (.5 mfd.)	30–4015
④	Vibrator	38–5036
⑤	Resistor (200 ohms)	7217
⑥	Resistor (200 ohms)	7217
⑦	Transformer	32–7030

No. on Fig. 3	Description	Part No.
⑧	Condenser (.006 mfd.)	30–1002
⑨	Condenser (4 mfd.; 8 mfd.)	30–4010
⑩	Filter Choke	32–7026
⑪	R. F. Choke (High voltage)	32–1078
	84 Tube	32–2001
	Battery Cable (Model 6F)	41–3017

PHILCO RADIO & TELEVISION CORP.

MODEL 5 ADJUSTMENTS

Become thoroughly familiar with the adjustment procedure and the location of the padding condensers before starting to adjust a Model 5 Receiver.

Furthermore, don't attempt to make the adjustments using a make-shift oscillator. The modern radio depends on critically tuned circuits for its exceptional performance. It is nothing short of gross carelessness to try to adjust these delicately tuned circuits using unstable oscillators which are incapable of being calibrated accurately.

Use a Philco 095 oscillator, or if your service department is fortunate enough to have one, the new Philco Signal Generator 048.

NOTE.—United Motors Service Stations, see U. M. S. Service Manual.

The intermediate frequency used is 460 K. C. Set up the oscillator or signal generator for this frequency.

Disconnect the grid lead from the 6A7 tube. Then connect the test lead to the grid of this tube and ground the shield on the Receiver housing. Use the fibre adjusting wrench 3164 for all adjustments.

Padder 10. Turn the adjusting nut in until tight. Then back off one full turn. Leave this condenser in this position until the last step.

Padder 11. This is the first I. F. primary condenser. With the Receiver and oscillator turned on and the oscillator set for 460 K. C., turn the Receiver volume control on full and adjust the oscillator attenuator. Then adjust the padder for maximum signal in the loud speaker.

Padder 13. This is the first I. F. secondary condenser. Adjust the attenuator so that the signal is barely audible. This should be repeated with each adjustment if necessary. Adjust the padder for maximum signal in the loud speaker. Repeat this procedure in the next two adjustments.

Padder 17. This is the second I. F. primary condenser.

Padder 20. This is the second I. F. secondary condenser.

Remove the oscillator connections from the 6A7 tube and reconnect the Receiver grid lead to this tube. The oscillator setting must now be changed to 1500 K. C.

The Receiver volume control must be turned on full, the oscillator lead connected to the antenna lead-in and the shield to the Receiver housing. To obtain the correct setting of the tuning condenser, open the plates as wide as possible. Place a piece of paper on the stator plates and then turn the rotor in until it strikes the paper.

Oscillator padder. This is the padder on the second section of the tuning condenser (section nearest drive mechanism). Adjust for maximum signal.

Antenna Padder. This is the remaining padder on the tuning condenser. Remove the paper from the tuning condenser and set the condenser and oscillator for 1400 K. C. Adjust the padder for maximum signal.

Low Frequency Padder 10. Set the oscillator for 600 K. C. and tune the Receiver to this frequency. Adjust the padder for maximum signal. After completing these operations, repad the antenna padder at 1400 K. C.

Model EG Vibrator—Continued from Page 1

The terminal panel provides for the following connections:

A ± terminal for control, connecting to the control relay.

+B terminal, 180 volts to 200 volts for the "B" lead to the Receiver.

INT+B terminal, an intermediate voltage for Receivers requiring a tap voltage.

—B terminal, for Receivers requiring this lead. Normally it is not grounded. This, however, can be accomplished by strapping to the GND terminal.

GND terminal for grounding the chassis.

Complete instructions for installing are packed with each Vibrator.

USING THE EA DYNAMOTOR

Many Dealers and Service Stations have built up a profitable business selling and installing the EA Dynamotor for replacing "B" batteries and other power devices. A bit skeptical at first, they soon realized the market for this dynamotor and since then, repeat orders have come in, in nice volume. Intended primarily for use with the Model 3 and Model 7 as a battery replacement, service men have been quick to adapt it to all other makes of battery operated car radio.

The installation instruction label is pasted to the inside bottom of the dynamotor housing, where it can be seen by anyone making the installation. It is vitally important that these instructions be carried out in detail.

Since the EA was first placed on the market, an additional filter condenser has been placed on the "B+" lead. This condenser, 3615-AZ, is mounted on the base at the rear of the dynamotor. When one of the EA dynamotors equipped with this condenser is installed with the Model 3 Philco Transitone or any radio in which "B—" is not grounded, this additional change must be made:

Remove the mounting screw from the 3516-AZ condenser. Bend up the ground terminal which normally is grounded by the mounting screw. Replace the mounting screw and be sure that the old ground terminal does not make contact with the screw. This is important.

The "B—" lead, the black lead coming from the rear of the dynamotor, which is connected to the ground terminal on the base, must be disconnected from the ground terminal and connected to the new terminal on the 3615-AZ.

The "B—" terminal on the condenser must then be connected to the "B—" terminal on the terminal panel. This was formerly the "B+" screen terminal.

This additional change must be made on all Model EA dynamotors having the 3615-AZ condenser connected to "B+" when using the dynamotor with a Model 3 or any other Receiver with a non-grounded "B—", otherwise it will be impossible to clear up the dynamotor hum.

PHILCO RADIO & TELEVISION CORP.

Model 6, B-6, 9 and 12 Improvements

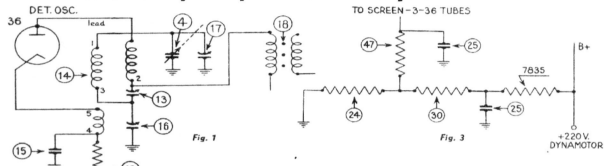

Fig. 1

Fig. 3

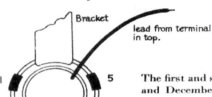

Fig. 2

Figures 1 and 2 show the schematic and terminal arrangements of the new oscillator coil in the Models 6, B-6, 9 and 12. Practically the same changes have been made in the 6 that were made in the 9. The dial becomes 8255 on the 6, and 8257 on the B-6. Padders 04000-D and 04000-A become 04000-X and 04000-J respectively. The 6000 ohm resistor 7352 becomes 8267 (13,000 ohms). The antenna transformer is now 06914; the R. F. Transformer 06915; the oscillator 06916 and the first I. F. Transformer 06932.

The first and second I. F. part numbers were transposed on the parts list in the November and December SERVICE BROADCAST. Number 05901 is first and 05970 is second. In the February SERVICE BROADCAST under Model 9 changes, the 8000 ohm resistor 8255 should be 7835.

Figure 3 shows the circuit arrangement of the screen voltage divider and filter circuit. Resistor 7835 has been added to the Models 9 and 12 only.

MODEL 12—122

The original Model 12 was similar to the Model 8 and was properly known as Model 12—Code 121. The present Model 12 is the Model 12—Code 122, and is similar to the Model 9 except that it is for 12 volt operation. The tubes, the circuit and the base arrangement are the same. Figure 4 shows the wiring of the tube heater circuits. Since 6.3 volt tubes are used, a series multiple connection must be used to operate them from a 12 volt battery.

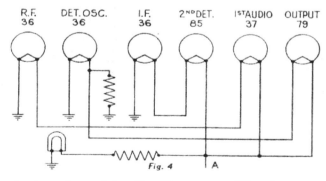

Fig. 4

The shunt resistor on the oscillator tube is Part No. 33-3002, 21 ohms. The pilot light resistor is Part No. 7155, 30 ohms. The speaker is the A-9 and is equipped with a 12 volt field. The Model EE dynamotor is used, supplying 40 milliamperes at 220 volts.

The Model 12 has been designed especially for bus and boat installations where 12 volt battery systems are used. It gives the same matchless performance as the Model 9 and is priced the same, $89.50.

PHILCO RADIO & TELEVISION CORP.

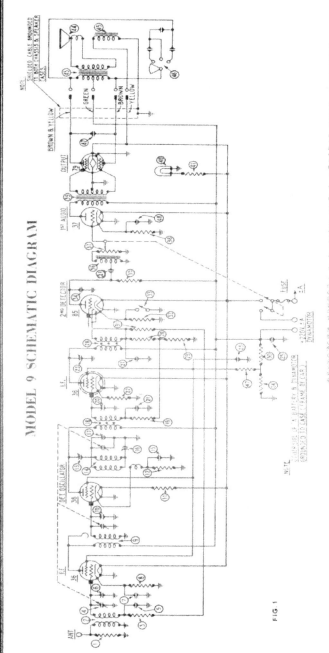

MODEL 9 SCHEMATIC DIAGRAM

FIG. 1

PARTS LIST

No. in Figs. 1 and 2	Description	Part No.
1	Resistor (5,000 ohm)	.6096
2	Antenna Coil	.06574
3	Resistor (100,000 ohm)	.6099
4	Tuning Condenser	.04308
5	By-pass Condenser (.05 mfd.)	.3615-AN
6	Compensator section on tuning condenser	..
7	By-pass Condenser	3615-AY
8	Resistor (500 ohm)	.6977
9	R. F. Transformer	.05992
10	Compensator section on tuning condenser	..
11	Resistor (2.7 ohm)	.6511
12	Resistor (6,000 ohm)	.7352
13	Compensator	.04000-A
14	Oscillator Coil	.05975
15	Condenser (.0007 mfd.)	.04520
16	Compensating Cond.	.04000-S
17	Compensator section on tuning condenser	..
18	First I. F. Transformer	.05970
19	Resistor (500,000 ohm)	.6097
20	Compensating Cond.	.04000-D
21	Condenser (.05 mfd.)	.06091
22	Resistor (500 ohm)	.6977
23	Compensating Cond.	.04000-D
24	Resistor (20,000 ohm)	.6650
25	Condenser (.5 mfd. .25 mfd.)	.06088
26	Second I. F. Transformer	.05901
27	Condenser (.00025 mfd.)	.3082
28	Resistor (100,000 ohm)	.6099
29	Resistor (100,000 ohm)	.6099
30	Resistor (20,000 ohm)	.6649
31	Resistor (500,000 ohm)	.6097
32	Resistor (5,000 ohm)	.6096
33	Switch	.5462
34	Condenser (.00025 mfd.)	..
35	Resistor (50,000 ohm)	.5886
36	Audio Transformer	.4518
37	Volume Control	.7525
38	Resistor (2,500 ohm)	.7775
39	Input Transformer	.7652
40	Pilot Lamp	.4567
41	Resistor (7 ohm)	.5510
42	Condenser (.06 mfd.)	.5259
43	Output Transformer	.2515

No. in Figs. 1 and 2	Description	Part No.
44	Speaker Coil and Cone	.02823
45	Speaker Field Pot.	.02795
46	Tone Control	.05366
47	Resistor (25,000 ohm)	.4549
48	Condenser	.7774
	Complete Speaker Assembly (Model 6)	.A-4
	Complete Speaker Assembly (Model 7)	.A-4
	Complete Speaker Assembly (Model 8)	.A-5
	Complete Speaker Assembly (Model 9)	.A-7
	Complete Speaker Assembly (Model 12)	.A-6
	Complete Speaker Assembly (Model B-6)	.A-8
	Interstage Shield	.05910
	Dynamotor ED	.06084
	Dynamotor EA (for battery replacements)	.05388
	Receiver Studs	.6122
	Shielded Loom (18" high tension shield)	.L-1387
	Shielded Loom (30" high tension shield)	.L-1386
	Spark Plug Resistor	.4531
	Distributor Resistor	.4546
	Screw Type Resistor	.4851
	Interference Condensers	.4522
	Knobs	.5166
	Speaker Extension Cables	.02984
	Dynamotor Filter Choke	.6658
	Dynamotor Filter Condenser (large unit)	.05586
	Dynamotor Filter Condenser (small unit)	.05724
	Dynamotor RF Choke	.5723
	Battery Cable	.05419-D
	18" Volume Control Shaft	.6351
	18" Tuning Control Shaft	.6352
	32" Volume Control Shaft	.6128
	32" Tuning Control Shaft	.6129
	48" Volume Control Shaft	.6298
	48" Tuning Control Shaft	.6299
	120" Volume Control Shaft	.6355
	120" Tuning Control Shaft	.6356
	Philco Oscillator (for adjusting Models 5, 6, 7, 8, 9)	.095
	Fibre Wrench	.3164

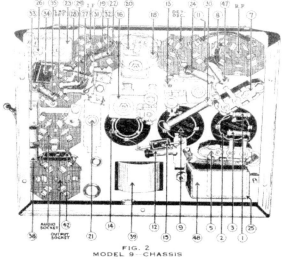

FIG. 2
MODEL 9—CHASSIS

PHILCO RADIO & TELEVISION CORP.

The New Model 6

Philco Transitone's latest five-tube all-electric super-heterodyne, the Model 6, is the finest automobile radio Receiver. Extreme sensitivity, remarkable selectivity, even finer tone than ever before, all-electric—and many other added features make this Receiver the outstanding automobile radio today.

The automatic volume control is more effective, due to the manner in which the new 85 type tube is used. This tube combines the action of a diode type detector with that of a triode amplifier. This arrangement gives perfect detector action and, in addition, a stage of audio amplification ahead of the 41 tube.

These improvements, coupled with the proven superiority of Philco's balanced unit construction guarantee maximum lasting consumer satisfaction with the minimum service trouble to you.

Philco Transitone's improved dynamotor, the Model EB, is furnished as standard equipment with the new Model 6 Receiver.

Be Sure You Know How To Do This

The intermediate frequency of the Model 6 is 260 K.C. This is a departure from the frequency used in the Model 7 and 8 Receivers. All dealers and installation stations must be equipped with a suitable oscillator capable of producing accurately a 175 K.C. signal for the Models 7 and 8 and 260 K.C. for the Model 6.

Philco's oscillator, Model 095, priced at $28.50 net to the dealers and service stations, is the ideal oscillator for such work and can be ordered direct from your distributor.

I. F. Stages

Remove the grid clip from the detector oscillator tube and connect the output of the oscillator to the control grid. The detector oscillator is the second tube from the right.

With the Receiver and oscillator turned "on," set the oscillator for 260 K.C. and adjust the oscillator attenuator so that the signal is barely audible with the Receiver volume control turned on full. If the oscillator is equipped with an output meter, connect the meter and adjust the attenuator so that a half scale reading is obtained.

Using a Philco 3164 fibre wrench, adjust the second I. F. condenser. This is numbered (23) on figs. 1 and 2. The correct adjustment is obtained when the strongest signal is heard in the speaker or the maximum reading is secured on the meter.

Next adjust the secondary and primary I. F. condensers. These are (20) and (13), respectively, on figs. 1 and 2.

Disconnect the oscillator and reconnect the clip to the control grid.

High Frequency Compensators

Connect the output of the oscillator to the antenna lead and the housing of the Receiver. With the Receiver turned on and the oscillator set for 175 K.C., tune the Receiver to 1400 K.C., the eighth harmonic of 175 K.C., and adjust the third padder on the tuning condenser for maximum signal. This is the one on the extreme left of the housing. The purpose of this adjustment is to line up the condenser so that 1400 K.C. is tuned in at 140 on the scale when the scale is set properly.

It may be necessary to adjust the first two compensators on the tuning condensers at 1400 K.C., in order to get a strong enough signal through.

R. F. Compensators

After the detector oscillator has been padded at 1400 K.C., adjust the first and second R. F. Condensers on tuning condenser at 1400 K.C.

Low Frequency Compensator

Now tune the Receiver to 700 K.C. and adjust the condenser (16) on figs. 1 and 2. During this operation the tuning condenser must be shifted and the compensator must be adjusted to bring in the maximum signal.

After this has been done, check the adjustment of the high frequency condenser at 1400 K.C. again.

PHILCO TRANSITONE DYNA-MOTORS

In all there are five dynamotor models with characteristic differences.
Model EA 6.3V. primary, 40 milliamperes at 180 V. (67½ V. tap.)
Model EB 6.3V. primary, 40 milliamperes at 180 V.
Model EC 12.6V. primary, 40 milliamperes at 180 V.
Model ED 6.3V. primary, 40 milliamperes at 220 V.
Model EE 12.6V. primary, 40 milliamperes at 220 V.

Standard Equipment

Philco Transitone Model 6—Dynamotor EB.
Philco Transitone Model B6—Dynamotor EB.
Philco Transitone Model 7—Dynamotor EA.
Philco Transitone Model 8—Dynamotor EA.
Philco Transitone Model 9—Dynamotor ED.
Philco Transitone Model B9—Dynamotor ED.
Philco Transitone Model 12 (121)—Dynamotor EC.
Philco Transitone Model 12 (122)—Dynamotor EE.
Packard Car Radio—Dynamotor ED.

PHILCO RADIO & TELEVISION CORP.

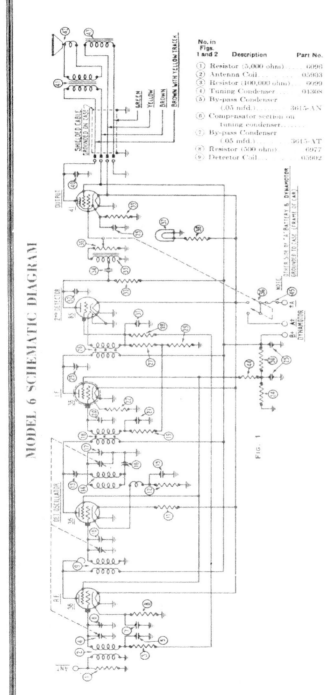

MODEL 6 SCHEMATIC DIAGRAM

FIG. 1

PARTS LIST

No. in Figs. 1 and 2	Description	Part No.
①	Resistor (5,000 ohm)	6096
②	Antenna Coil	05903
③	Resistor (100,000 ohm)	6099
④	Tuning Condenser	04368
⑤	By-pass Condenser (.05 mfd.)	3615-AN
⑥	Compensator section on tuning condenser	
⑦	By-pass Condenser (.05 mfd.)	3615-AT
⑧	Resistor (500 ohm)	6977
⑨	Detector Coil	05902

No. in Figs. 1 and 2	Description	Part No.
⑩	Compensator section on tuning condenser	
⑪	Resistor (2.7 ohm)	6511
⑫	Resistor (8,000 ohm)	5838
⑬	Compensating Cond.	04000-A
⑭	Oscillator Coil	05975
⑮	Condenser (.007 mfd.)	4520
⑯	Compensating Cond.	04000-S
⑰	Compensator section on tuning condenser	
⑱	First I. F. Transformer	05970
⑲	Resistor (500,000 ohm)	6097
⑳	Compensating Cond.	04000-D
㉑	Condenser (.05 mfd.)	3615-AK
㉒	Resistor (500 ohm)	6977
㉓	Compensating Cond.	04000-D
㉔	Resistor (20,000 ohm)	6650
㉕	Condenser (25 mfd., 5 mfd., 8 mfd.)	04354
㉖	Second I. F. Transformer	05901
㉗	Resistor (100,000 ohm)	6099
㉘	Resistor (500,000 ohm)	6097
㉙	Resistor (100,000 ohm)	6099
㉚	Resistor (20,000 ohm)	6649
㉛	Condenser (.00025 mfd.)	3082
㉜	Condenser (.0002 mfd.)	4059
㉝	Resistor (50,000 ohm)	4237
㉞	Condenser (.09 mfd.)	4989-Y
㉟	Audio Transformer	7535
㊱	Volume Control (500,000 ohm and switch)	7525
㊲	Pilot Lamp	4567

No. in Figs. 1 and 2	Description	Part No.
㊳	Resistor (7 ohm)	5110
㊴	Resistor (700 ohm)	6443
㊵	Condenser (.002 mfd.)	6853
㊶	Output Transformer	2598
㊷	Cone and Coil	02823
㊸	Field Coil	02794
㊹	Resistor (25,000 ohm)	4516
	Interstage Shield	05910
	Dynamotor EB	05389
	Dynamotor EA (for battery replacements)	05388
	Receiver Studs	6422
	Shielded Loom (18" high tension shield)	L1387
	Shielded Loom (30" high tension shield)	L1386
	Spark Plug Resistor	4531
	Distributor Resistor	4546
	Screw Type Resistors	4534
	Interference Condensers	4522
	Knobs	5166
	Speaker Extension Cable	02984
	Dynamotor Filter Choke	6658
	Dynamotor Filter Condenser (large unit)	0538
	Dynamotor Filter Condenser (small unit)	05724
	Dynamotor RF Choke (small unit only)	05746
	18" Volume Control Shaft	6351
	18" Tuning Control Shaft	6352
	32" Volume Control Shaft	6128
	32" Tuning Control Shaft	6129
	48" Volume Control Shaft	6298
	48" Tuning Control Shaft	6299
	120" Volume Control Shaft	6355
	120" Tuning Control Shaft	6356
	Philco Oscillator (for adjusting Models 3, 7, 8, 6)	Model 095
	Fibre Wrench	3164

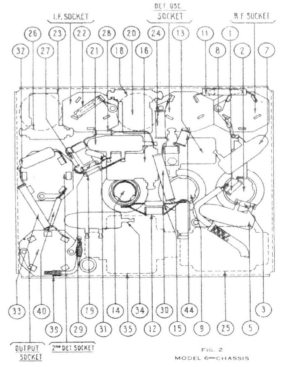

FIG. 2
MODEL 6—CHASSIS

PHILCO RADIO & TELEVISION CORP.

A RADIO INTERFERENCE TROUBLE FINDER

THERE are many Transitone Receivers being used by power companies and radio service men for locating the sources of power leaks and radio interference. While most of these Receivers are being used just as they were received from the factory, there have been numerous requests for information as to the best method of removing the automatic volume control feature from the Receivers. Others have desired a switching arrangement, appreciating the value of automatic volume control for broadcast reception, but realizing that without it, the Receiver will pick up more interference.

In Fig. 1 the I. F. and detector stages of the Model 9

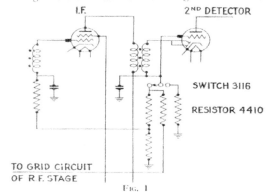

Fig. 1

are depicted with the switch for controlling automatic volume control. Similar changes can be made to any Philco Transitone Receiver since all models use the same automatic volume control principle. The single pole double throw switch in the normal position connects the R. F. and I. F. grid circuits and the resistor network to the second detector stage. In the other position, these circuits are disconnected from the second detector while the detector stage is terminated with another resistor.

The connections can be broken at the by-pass condenser ⑰ and the switch connections made to the condenser and resistors at this point.

For best operation in locating interference, a small enclosed loop should be used. This can be conveniently mounted on a pipe stand on the left running board forward of the front corner post. The upper section of the loop should be free to rotate so that it can be turned

toward the source of interference and in this manner be easily located.

With automatic volume control connected in the circuit, it is difficult to get a good location of the source of interference because of the action of the automatic control in holding the output of the Receiver to the same volume level over such a wide range of varying signal strength. With this control cut out, the least change in signal strength will be noticeable in the Receiver output.

Radio dealers and servicemen can use a Transitone fitted out as described above to such good advantage in clearing up chronic cases of interference in their respective territories, thereby increasing their list of prospects for the better modern home Receivers. In addition, it better qualifies them to sell interference elimination to the owners of this noise-producing equipment, opening up a new field to most of them.

INSTALLATION OF THE MODEL 5 IN MODEL "A" FORDS

THE rigid choke rod in the Model "A" Ford prevents installing the Model 5 on the right-hand side of the dash, and there is not sufficient room on the left side above the steering column.

To overcome this condition, remove the rigid choke rod and install the flexible choke used in the Model "B" Fords. The parts necessary for the change are 1—B-9700, 1—B-9709 and 1—B-9570. These can be obtained from the local Ford dealer's service department and will cost about 57 cents. The operation is very simple, taking only a few minutes to make the change.

Remove the old rigid choke rod and the connector sleeve fitting from the carbureter. Save the spring and washers. These are to be used over again. The old mounting bracket can be reused for the new choke rod. Dress the flexible rod over the gas line and through the hole in the dash. Slide the washers and the long spring over the flexible rod casing. Then push the new spring over the casing and slide the new connector sleeve in place. The small spring ring must be squeezed in place in the groove in the end of the choke and then the half-round end seated in the carbureter.

The procedure sounds complicated, but the entire operation can be done in less than ten minutes.

With the flexible choke rod permanently installed, the Model 5 can then be conveniently located on the right-hand side of the dash.

Auto Radio Hints

There are still a few installation men who have an occasional job on which they cannot get rid of all motor noise. Usually this is caused by the distributor. The high tension terminals or contacts are not lined up perfectly due to shrinkage or warping of the head or to wear in the distributor gears. The rotor may strike a few of the contacts and miss the rest.

When peening a rotor under such conditions, the best plan is to chalk the contacts or terminals and then after the rotor has been carefully peened, turn over the motor a few times with the ignition turned off. Remove the distributor head and examine the chalked terminals. If the rotor has cut the chalk on a few of the contacts, these contacts should be scraped down with a hard sharp tool and the rotor again peened.

This procedure should be carried on until the rotor just traces a line through the heavy chalk layer on all the contacts. Obstinate cases of interference can be eliminated this way.

PHILCO RADIO & TELEVISION CORP.

"TRANSITONE MODEL 3"

Automotive Battery-Operated Receiver with Automatic Volume Control)

This model, manufactured by Transitone Automobile Radio Corporation, Philadelphia, Pa., bears no resemblance to previous "Transitone" models described in past issues of RADIO CRAFT.

Of exceptional interest is the inclusion of automatic volume control; a two-element or diode detector is used. "C" bias is obtained by resistor-drop, as in socket-power sets, as the schematic circuit indicates, there are 20 resistors in this battery-model receiver.

The values of the various components are as follows: resistor R1, 10,000 ohms; R2, R7, R8, R13, 0.1-meg.; R3, R4, R6, 250 ohms; R5, R12, R20, 1. meg.; R9, R10, 30 ohms; R11, R15, 0.25-meg.; R14, volume control; R16, 25,000 ohms; R17, 50,000 ohms; R18, 500 ohms; R19, 300 ohms.

Condensers C1, C2, C3 are the usual tuning units; C4, C5, C6, C9, C10, .05-mf.; C7, C18, 1 mf.; C8, C16, 0.25 mf.; C11, C13, C14, .00025-mf.; C12, .0005 mf., C15, .015-mf.; C17, 2 mf.

Resistors R6, R9, R10 and R20 are contained in one unit; and resistors R18 and R19 in another. Resistors R3 and R4 are combined with condensers C9 and C10.

It should be obvious that the most important single factor in correct operation of this model receiver, aside from tubes of correct characteristic, is the use of resistors of correct constants. The wattage ratings of the resistors are as follows: R1, R5, R7, R8, R11, R13, R15, 0.5-watt; R2, R12, R16, R17, 1 watt. The resistor color code is as follows: R1, black; R2, R7, R8, R13, silver gray, yellow tip; R5, R12, green, white tip; R11, R15, white; R16, brown, yellow tip; R17, orange; R18-R19, and R20, flat wire-wound.

Tube average operating characteristics are as follows: filament potentials, V1, V2, V3, 2 volts; V4, V5, V6, V7, 5 volts. Plate potentials: V1, V2, V3, 150 volts; V4, zero; V5, 45 volts; V6, 140 volts; V7, 142 volts. Control-grid potentials (negative): V5, 1.0 volt; V6, 2.5 .volts; V7, 32 volts. Cathode potentials: V1, V2, V3, 2 volts. Screen-grid potentials: V1, V2, V3, 80 volts. Plate currents: V1, V2, V3, 1.5 ma.; V4, zero; V5, 1.0 ma.; V6, 3 ma.; V7, 16 ma.

If it becomes necessary to re-align the tuned circuits to obtain greater selectivity and volume, use a fiber wrench and adjust the trimmers for a signal between 1,000 and 1,200 kc.; starting first at C3.

Noisy operation may be due to a poor bond between the receiver chassis and the car chassis. A partial test for this possible source of trouble is to remove the antenna leads when noise due to this cause will continue unabated.

Lack of sensitivity, or noisy operation, may be due to close proximity of the antenna in the top of the car to the metal-work, the a(??) should be spaced from all such conductors (??) instance, the dome light) by a distance of (??) least 3 inches.

There is only one "A" lead; it is black, and terminates in a lug. Connect this to one of the ammeter terminals on the instrument panel, so that the current drain of the radio set does not show on the meter. The charging rate of the car storage battery should be increased about 2 amps., to compensate for the average amount of current consumed by the radio set.

After servicing an automotive receiver it is important to see that all metal parts—shielding, cable sheaths, etc.—are well grounded to the chassis of the car. Tubes and batteries after replacement must be securely fastened in place.

If it becomes necessary to replace the flexible tuning shaft, the procedure is as follows: push the free end of the flexible shaft through the bracket on the receiver so that the tip of the shaft is seated in the coupler. Tighten the two set-screws on the coupler, and then tighten the set-screw on the bracket just enough to hold the casing in place. Tune in a station of known frequency, adjusting the receiver exactly. Loosen the two set-screws on the coupler which lock the shaft in place. The flexible tuning shaft can then be turned without affecting the setting of the tuning condenser in the receiver. Set the dial scale accurately to the channel number corresponding to the station frequency, and re-tighten the two set screws on the coupler. Check at several points the relation between dial reading and station frequency.

The best material for an aerial is No. 14 or 16 copper screening, 36 in. wide. It should be used to replace all galvanized iron poultry-screen, where the twisted parts are not bonded; cutting and lacing back the latter to make room for the copper screen. Most car tops are of wooden-bow and cloth construction, with perhaps poultry-screen; but, where steel bows are used, instead, greater sensitivity sometimes is obtained by lacing in an antenna of stranded rubber-covered wire.

Poor tone quality may be due to an air space between the reproducer and the baffle (Part No. 2697-A) which should be used with it.

Standard interference suppression includes the use of standard spark-plug series resistors, a distributor (high-tension-lead) series resistor, and interference bypass condensers on the brush side of the generator cutout, and the battery or ammeter side of the ignition coil.

If this procedure (described in detail in past issues of RADIO-CRAFT magazine) does not result in sufficient suppression, it may be necessary to try the following: move the ignition coil from inside of dash to engine side of partition; shield the high- and low-tension leads from the ignition coil to the dash; and securely ground the shielding, or mount the coil on the engine side of the dash. (In some instances the construction of coil and switch may render this impossible; when it will be necessary to use a separate coil and mount it in the engine

compartment). Note particularly that only in rare instances should high-tension leads be shielded; for which purpose "shielded high-tension cable" must not be used.

It may be necessary, in some cases, to connect the "A—" black-with-white lead to the battery instead of the battery-side of the ammeter; and perhaps shield the lead, grounding the shield (copper braid over loom) in several places—a procedure which is particularly efficacious. Improved reception then indicates that further correction should be applied: shielding of the speaker cable, and the battery cable between set and control-unit. All shielding should be grounded. (Commercial shielded-cable is preferable to separate shielding.) In some cases it is desirable to shield the lead from antenna to set; using only "shielded high-tension cable."

Interference due to dome-light coupling may be eliminated by connecting bypass condensers where these wires enter the corner post. Dirty distributor contacts may cause noisy operation; over-wide separation of its contacts may cause the same effect. Reversing the ignition coil's primary leads sometimes reduces interference. Rubbing metal parts of the car chassis occasionally require bonding to the body of the car to reduce crackling sounds; cables, rods and pipes unless grounded may act as ignition-noise carriers. Pay particular attention to the temperature-indicator tube and the oil lines.

Fender, seat, and door pads are available, for use to prevent marring the finish of a car when installing or servicing the radio installation.

Dome-light and switch wiring must be run along the side of the top frame, and along the top edge of the side of a bow to the dome-light fixture.

Lack of signals, or weak signals, may be an indication of a grounded antenna.

All conductors should be well insulated from the car chassis, to prevent short-circuit; while fuses in the "B—" and "A" leads adds a safety factor.

It is suggested that a complaint of poor service from the "B" batteries may be checked by reference to the speedometer's mileage indication for the period of the installation of the batteries. This figure, divided by 25, gives approximately the number of hours the radio set has been used; which, divided by the figure for the elapsed time, in days, since the installation of the batteries, indicates the number of hours per day the radio set has been in use. Heavy-duty "B" blocks should last about 600 operating days (1 hour per day), to 150 days (4 hours per day).

The distributor rotor should just clear all stator contacts (test chalk marks on these contacts should remain undisturbed); file the contacts; or file or peen the rotor, as may be required.

Credit for these data is hereby extended to Messrs. Robert F. Herry and Robert Long, Jr., of the manufacturer's service department.

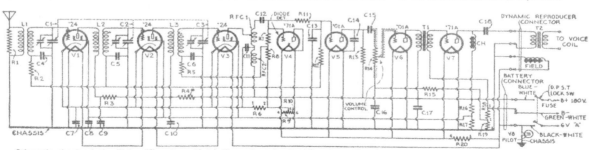

Schematic circuit of the Philco "Transitone Model 3" receiver, incorporating automatic volume control, a necessity in automotive radio sets to overcome the effects of changing location; the total current consumption is 4 amps. The reproducer is catalogued as the "Transitone Model 3 Dynamic Loud Speaker." Resistor R14 is of standard 0.5-meg. rating.

PIERCE AIRO DEWALD MODEL 52

PIERCE - AIRO, INC.

AN interesting design feature of the Pierce Airo DeWald model 52 Motortone receiver is the inclusion of the receiver chassis, dynamic reproducer, and the "B" eliminator in one complete assembly, as shown in the illustration. An interference suppression kit is furnished with the equipment. The type and arrangement of the antenna is optional with the Service Man. Since most modern cars are factory-equipped with an antenna, this very necessary signal pick-up unit is not included in the standard Motortone offered by Pierce Airo.

The receiver assembly is designed to be mounted on the rear dash under the cowl; just run two leads for the "A" and a third for the antenna, and the job is done. There is supplied with the equipment a template of the four mounting-stud holes. The illuminated remote control unit clamps to the steering post. When inserting the tubes in the set, the type 37 tube must be inserted before the type 89 tube.

The superheterodyne circuit used in this set has an I.F. of 175 kc. An A.V.C. connection is incorporated. Tubes utilized: One type 36 tube as combined oscillator and first-detector; one 36, I.F. amplifier; one 85, A.V.C. and second-detector; one 37, first A.F.; and one 89, second A.F.

The "B" unit built into the assembly is of the interrupter type; therefore, do not remove the tubes or the cable plug while the set is on. The "B" rectifier is a type P861.

PREMIER "AUTO PAL" RECEIVER

A STANDARD kit of parts, illustrated, for the "Auto Pal" set includes the following items: a 5-tube superheterodyne receiver, tubes, separate dynamic reproducer, steering-column control unit, flexible antenna pad, kit of interference suppressors, distributor filter condenser, and miscellaneous hardware. Individual car designs determine the most suitable place to install this equipment. Provision must be made for the "B" supply, which is not included in the standard kit of parts; the "B" voltage may be obtained from batteries or an available Premier "B" eliminator accessory rated at 30 milliamperes and 180 volts.

In this superheterodyne chassis an I.F. of 175 kc. is used. The following tubes are supplied by the manufacturer: One type 87, oscillator-first-detector; one 88, I.F. amplifier; one 75, second-detector and A.V.C.; one 37, first A.F.; one 89, second A.F. The total power consumption at 6.3 volts battery voltage is 33.5 watts. The sensitivity is 2 microvolts-per-meter; the power output, 1.5 watts.

Test voltages for this set are given in the following table; all values are to ground (chassis) and at zero signal input; filament potential, 6.2 volts:

Tube	Tube Position	Cath. Volts	S.G. Volts	Plate Volts
87	Osc.-Det. 1	2	70	150
88	I.F.	4	70	165
75	A.V.C.-Det. 2	2		150
37	A.F. 1	15		180
89	A.F. 2	17	180	180

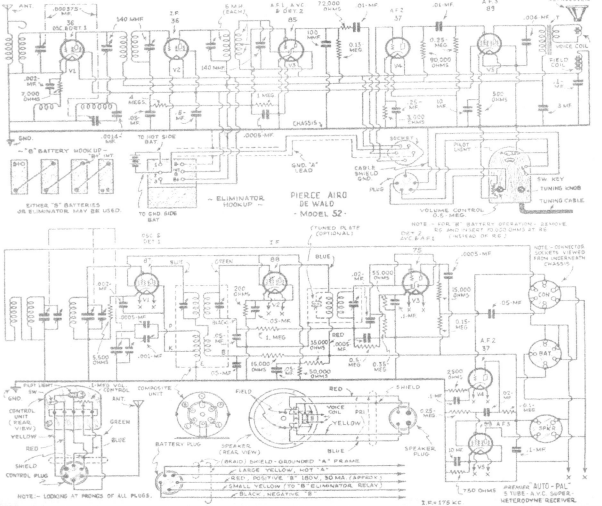

PIERCE AIRO DE WALD - MODEL 52 -

PREMIER "AUTO-PAL" 5 TUBE - A.V.C. SUPER-HETERODYNE RECEIVER

I.F. = 175 KC.

PIERCE - AIRO, INC.

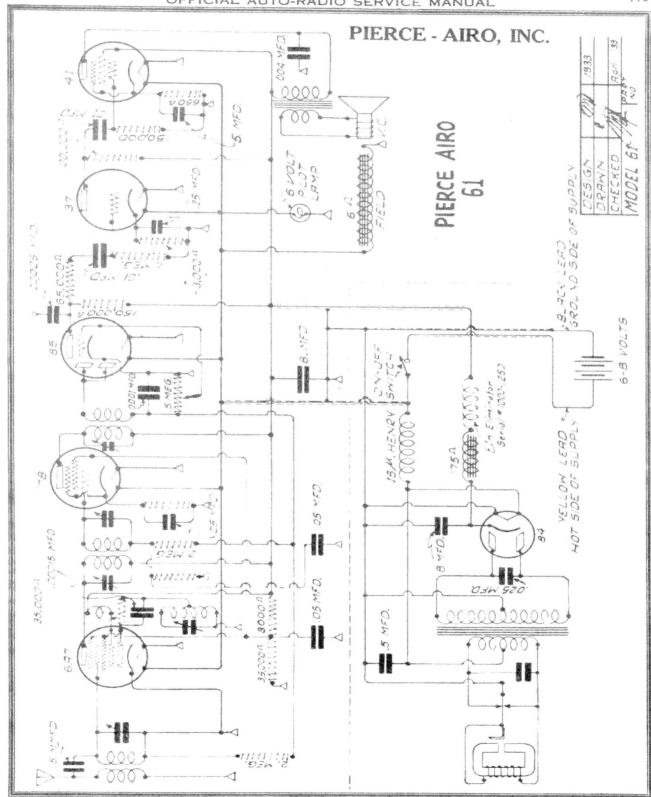

PIERCE AIRO
61

MODEL 61

RCA-VICTOR, Inc.

Instructions for

RCA Victor M-34

Automobile Receiver

INTRODUCTION

Mechanical simplicity and high-quality performance are keynotes of this automobile radio receiver. The instrument consists of a superheterodyne chassis, a loudspeaker, and a vibrator-type "B" battery eliminator mounted in a single case. It is operated entirely from the car storage battery.

A remote control unit, mounted on the steering column and connected to the receiver through a flexible shaft and cable, places all controls convenient to the driver. This unit contains the station selector control, a glare-proof illuminated dial (calibrated in station channels) and a combined volume control and "key-lock" power switch.

Equipment for the suppression of ignition interference is provided. The use of a roof (built-in or interior type) antenna is recommended.

PART I—INSTALLATION
Procedure

1. Unpack the set from carton and check equipment. (See "Equipment Furnished"

2. Remove tube packing inside receiver case and examine tubes. (See details under "Mounting of Units"   *Do not replace case cover.*

3. **CHECK POLARITY OF AUTOMOBILE STORAGE BATTERY SUPPLY.** If the negative (—) side is grounded to car frame, make changes to chassis connections shown in Figure 1. *Do not disturb these connections if positive (+) side is grounded.* (See details under "Mounting of Units"— Replace case cover.

4. Determine most satisfactory mounting position (see details under "Location of Units" spot mounting-bolt location and drill ½" diameter hole. Insert bolt through dash and assemble support plate and nuts on engine side. Hang receiver over bolt head and tighten nuts. (See Figure 1 and details under "Mounting of Units"

5. Attach remote control unit to steering column by means of mounting bracket and strap. (See Figure 1 and details under "Mounting of Units"

6. Assemble flexible shaft to receiver and remote control unit. (See Figure 1 and details under "Mounting of Units"

7. Connect metal-shielded lead from receiver to antenna by means of coupling connector. (See notes on antennas under "Location of Units"   and details of lead-in under "Connections"

8. Connect terminal at end of *black* lead from cable to binding-post of automobile ammeter (see Figure 1 and details under "Connections"   ). The ignition by-pass capacitor (equipped with two leads) should be installed at this time. (See Figure 1 and paragraph 4 under "Suppression of Ignition Interference"

9. Install spark-plug and distributor suppressors; also generator by-pass capacitor (see Figure 1 and paragraphs 1, 2 and 3 under "Suppression of Ignition Interference"

10. Push knob over shaft protruding through front of remote control unit. Observing the dial scale, rotate knob slowly—first to stop position slightly beyond "150" and then reverse to other stop position slightly beyond "55."

11. Insert key in lock on remote control unit and turn to extreme clockwise position. Dial should become illuminated immediately but the tubes will not reach proper operating temperature until after approximately 45 seconds. (See details under **"PART II—OPERATION"** and **"PART III— MAINTENANCE."**)

RCA-VICTOR, Inc.

Equipment

A. Equipment Furnished:

1. *Receiver Package*—Includes the receiver and remote control units joined by the wiring cable:

(a) The receiver contains one each of the following Radiotrons installed in sockets: RCA-78, RCA-6A7, RCA-6B7, RCA-89.

(b) The remote control unit contains one dial lamp (6-8 volts).

(c) The wiring cable includes one fuse (20 amperes) installed in attached fuse receptacle.

2. *Outfit Package*—Containing:

(a) Flexible shaft (33⅞ inches long).

(b) Receiver unit mounting bolt (⅞ inch diameter), dash support plate, and nuts (2).

(c) Steering column bracket for remote control unit with strap, screws (2) and lockwasher (1).

(d) Shield clamp for antenna lead-in wire with screw (1), lockwasher (1) and nut (1).

(e) Key (1) and knob (1) for remote control unit and eyelets (2) for antenna connector packed in small envelope.

(f) Ignition Interference Suppression Equipment:

6 Spark plug type suppressors (additional obtainable from your dealer).

1 Distributor type suppressor.

2 Capacitors.

(g) Instruction Book.

B. Additional Equipment Required:

1. *Antenna*—One of the following types:

(a) Roof (built-in) type—recommended.

(b) Roof (interior) type for attachment to head-lining inside car—also recommended. A special antenna of this type complete with pin-hooks and lead-in wire may be purchased from your dealer.

(c) Plate (sub-mounted) type for attachment to channel members of car chassis—alternative. An efficient plate antenna completely equipped for mounting and a specially-designed shielded lead-in wire also are obtainable from the dealer.

Location of Units

Receiver and Remote Control Units—The arrangement of units shown in Figure 1 is recommended and will be found applicable to the majority of automobiles. Consideration should be given to the possibility of interference of the receiver with other equipment beneath the instrument panel or of the mounting bolt with apparatus on the engine side of the dash. By placing the receiver unit toward the right-hand side of the dash, the flexible shaft will be of correct length as furnished in practically all cases. This position, however, may be considered impractical because of its universal preference for heating devices, necessitating installation of the receiver unit either near the center or at the extreme left-hand side of the dash and the use of a shorter flexible shaft. In such cases, the shaft may be either short-

ened (as described under "Mounting of Units") or exchanged for one of proper length by the dealer.

NOTE—Two support brackets are attached to the receiver case, one on the rear surface and the other on the right-hand side viewing the loud-speaker opening. The side bracket must be used when the unit is mounted at the extreme left-hand end of the dash in order to avoid sharp bends in the flexible shaft and resultant unsatisfactory operation.

As furnished, the remote control unit is equipped for attachment to the steering column of the car. Its clamp bracket is so designed that the driver may select from a wide variety of possible mounting positions for maximum accessibility. The associated bracket strap will be found to accommodate practically any diameter steering column. If considered desirable, however, the remote control unit may be supported upon the instrument panel by means of an accessory bracket procurable from the dealer.

Antenna:

(a) *Roof (Built-in) Type*—Best results will be obtained by use of a built-in roof antenna. The majority of modern automobiles (closed body types only) are already equipped with such an antenna installed at the factory, the lead-in wire from which will usually be found coiled up beneath the instru-

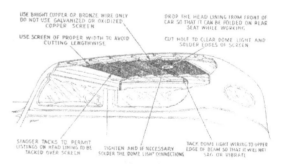

Figure 2

ment panel. Many other earlier cars employ a piece of metallic screen—for top material support—which, if ungrounded (not in electrical contact with the metallic frame), may be readily utilized as an antenna.

NOTE—The presence of a top support screen and of grounds in that screen may be determined without removing any portion of the inside fabric (head-lining). First procure any sharp-pointed metallic tool, push the point through the fabric (at several points if necessary) and feel around in an attempt to scrape the screen surface—being careful not to puncture the weather-proof top. If a screen is found, connect an ordinary dash or head-lamp between either terminal of the automobile ammeter and the tool, re-insert the tool through the head-lining and make contact with the screen. If the lamp lights, however dimly, it shall be assumed that the screen is grounded.

RCA-VICTOR, Inc.

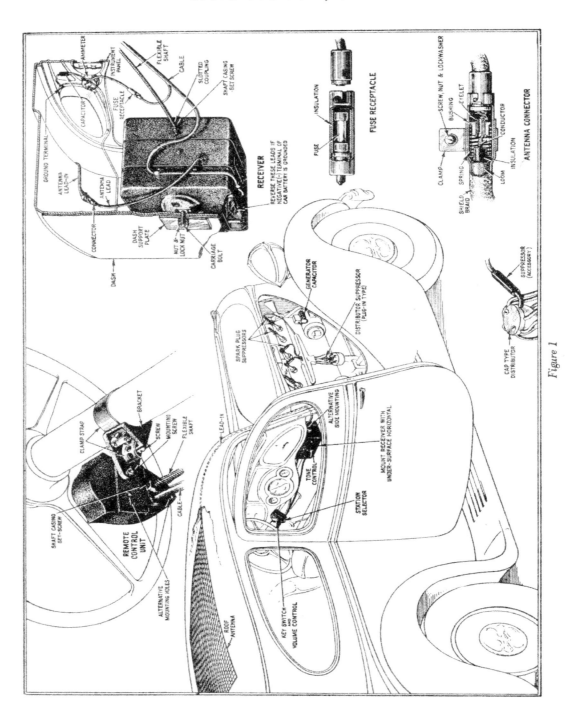

Figure 1

RCA-VICTOR, Inc.

In order to use an ungrounded support screen, first release the head-lining at the front corner nearest the receiver. Then connect a flexible rubber-insulated lead to the corner of the screen and solder the joint. Feed the free end of the lead down the adjacent pillar-post of the car into the driving compartment and replace the head-lining.

If the top support screen is grounded, or if no screen is present, it will be necessary to drop the entire head-lining (see Figure 2). In the former case, the screen may be insulated by removal of a strip several inches from all edges and from the dome light fixture. The possibility of subsequent shifting may be eliminated by tacking the screen to one or more of the ribs and by lacing the sides with cord. Where no support screen is used, a copper screen having a total area of at least ten square feet should be inserted. It should be located as far to the rear as possible and insulated from all metallic parts grounded to the frame of the car. The antenna finally should be tested for grounds (see the foregoing "NOTE" for test procedure). If satisfactory, attach the lead-in wire and replace the head-lining of the car.

NOTE—Since a degree of skill—only acquired by experience—is necessary in removing and replacing the top fabric material, such work should be allotted to a competent "trim" man.

(b) *Roof (Interior) Type*—The accessory interior-type roof antenna also will provide very satisfactory performance and, in addition, is extremely simple to install. It may be quickly attached to the head-lining inside the car (preferably as far to the rear as possible) by means of pin-hooks, thereby precluding removal of the fabric. An antenna of this type, however, should not be used in any automobile having *a grounded* top material support screen since the proximity of that screen would seriously reduce its efficiency. Before purchase, therefore, it will be advisable to check this possibility, following the test procedure described under *"Roof (Built-in) Type."*

As furnished, the interior-type antenna is equipped with a sufficient length of lead-in wire ready-attached. The effective antenna wire is enclosed by long-wearing paper procurable either in "gray" or "tan" finish as desired to harmonize with the car upholstery.

(c) *Plate Type*—For those cases where the installation of a built-in roof antenna is considered too costly and the interior roof antenna impractical, good reception from local or semi-distant powerful stations may be procured with the special plate-type antenna also obtainable as an accessory. This unit should be clamped to the frame of the chassis as far to the rear as possible. It is adjustable in length and may be mounted either lengthwise or crosswise of the chassis, which position should be selected with due regard to the prevention of overcrowding. The plate must be placed as close to the ground as possible, but not below the lowest portion of the chassis at the desired location, as sufficient road clearance must be retained. It is also important to avoid any position in which the plate will impede free motion of chassis parts such as springs, drive shaft, or axles in order to prevent damage to the antenna.

Mounting of Units

Details of mounting the various units are shown in Figure 1. The following procedures are recommended:

Receiver Unit—The rear cover of the receiver unit case (held in place by six screws) must be removed and all packing material (inserted for protection of the Radiotrons during shipment) withdrawn. Make certain that all Radiotrons are in the proper sockets and that the control grid clips are pressed down firmly over the respective dome terminals as shown by the diagram printed on the label affixed to the inside of the cover.

NOTE—At this point, it will be advisable to determine the electrical polarity of the storage battery supply. This may be done most conveniently by making an examination of the battery connections and ascertaining which terminal is grounded (that is, connected to the frame of the car). The positive terminal is usually marked (+) and tends to form corrosion far more rapidly than the negative (—). If the positive terminal is grounded, no change in the electrical connections of the receiver unit will be required. However, if the opposite is true, the two leads (equipped with spade terminals) located beneath the radio chassis as shown in Figure 1 must be reversed.

Now replace the rear cover and support the assembled unit against the dash in the chosen location. Allowing a clearance of at least two inches above the top surface, where possible, to permit subsequent removal of the case from the mounting bolt head, mark with a pencil or crayon on the dash four points corresponding to the corners of the adjacent case surface. Then determine the exact center of the area bounded by those four points (by drawing diagonal lines between opposite corners) and mark that position with a center-punch. Next drill a ½ inch hole at the center-punch mark and insert the mounting bolt. The support plate and the two nuts then should be assembled upon the bolt from the engine side of the dash as shown but should not be tightened. Finally hang the receiver over the bolt head, align sides vertically and tighten the nuts in place.

Remote Control Unit—In attaching the remote control unit to the steering column of the car, it will be advisable first to examine the detailed view (in Figure 1) showing the assembly of its mounting bracket. Four small holes are contained in the associated flexible strap at distances proper for use with steering columns of the most common diameters (1½, 1⅝, 1¾, 1⅞ inches) but the strap length will be found sufficient to permit the insertion of an additional hole if necessary to accommodate a 2 inch column. The proper hole may be determined by wrapping the clamp strap tightly around the column, inserting the machine screw furnished through that hole found to be nearest in alignment with the tapped hole in the clamp bracket. Three tapped holes are provided in the back of the remote control unit, permitting support of that unit either at the right- or left-hand side or above the steering column.

RCA-VICTOR, Inc.

Flexible Shaft—Insert that end of the flexible shaft to which is attached the slotted coupling through the bushed opening in the left side of the receiver unit. Then rotate the shaft from the free end until the coupling slot is felt to engage over the pin contained in the tuning mechanism and slide the shaft forward to the full depth of the slot. With the shaft held in this position, insert the opposite end of the shaft through the bushing at the rear of the remote control unit and push forward until the flatted portion of the shaft protrudes through the front cover. Then proceed to tighten the external set-screw (located at the bottom of the case—see Figure 3) adjusting the shaft position as necessary until the screw is felt to engage in the groove. Tighten the screw fully to the bottom of the slot and then loosen it approximately one-quarter of a turn. Finally, secure the flexible casing in place by tightening the set-screws at each end.

shielded and cut to eliminate excessive slack when attached to the receiver antenna connector. Before connecting the antenna to the receiver, the following comments applying to the particular type of antenna adopted should be observed:

(a) *Roof Antenna (Built-in Type)*—The lead-in wire from a factory-installed built-in roof antenna usually is unshielded and often is of insufficient length to reach the receiver. If necessary, an extra length of insulated wire may be spliced to the existing lead-in, in which case the joint must be soldered and wrapped with tape. In general, it will be advisable to shield the exposed length of lead-in wire, procuring for this purpose from your dealer a length of shield braid and an equivalent length of insulating loom (or rubber tubing) sufficient to extend between the end of the lead-in wire and its point of entrance from

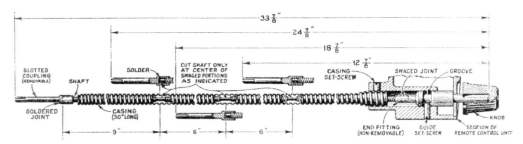

Figure 3

NOTE—In many installations it will be found necessary or desirable to use a flexible shaft of shorter length than 33⅞ inches. While it is simplest to procure a shaft of proper length from the dealer as mentioned heretofore, very little difficulty should be experienced in shortening the original part if deemed expedient. To shorten the shaft, refer to Figure 3 and proceed as follows:

1. Determine the minimum shaft length permissible for the installation.

2. Remove the slotted coupling (using a soldering iron) and withdraw the shaft from its casing.

3. Cut the shaft only at the center of a swaged joint, selecting that joint which allows at least the required length.

4. Cut from the shaft casing a length equal to the amount of shaft removed. (This operation may be simplified by placing the casing between wooden blocks in a vice so that the block ends will serve to guide the back saw blade.)

5. Replace the shaft in its casing and solder the slotted coupling to the end of the shaft.

Connections

Refer to Figure 1 and make connections as follows:

Antenna to Receiver—For least ignition interference, any portion of the antenna lead-in wire which extends behind the instrument panel or into the engine compartment of the car should be fully

the body pillar post. Slip the loom over the lead-in wire and the shield braid over the loom.

(b) *Roof Antenna (Interior Type)*—If an interior type antenna is used, the lead-in wire should be brought down the outside of that front pillar post nearest the receiver.

(c) *Plate Type Antenna*—With the plate type antenna, the fully-shielded end of the special cable should be brought into the automobile driving compartment through a ½ inch hole drilled in the toe-board (if no other opening is available). This end is to be connected to the receiver unit antenna lead (as explained in following paragraphs) and the opposite (unshielded) end then cut off as required to eliminate excessive slack upon connection to the plate. The pigtail extension from the end of the shield must be soldered or bonded to the frame of the car.

Refer to the detailed view of the antenna connector shown in Figure 1 and proceed to attach the lead-in wire (if shielded) as follows: First, cut the end of the lead-in so that the internal insulated wire and loom (if present) are flush with the end of the shield covering and push back the shield approximately 1½ inches. Cut the loom to the end of the

RCA-VICTOR, Inc.

shield and then remove sufficient insulation to expose one inch of clean bare-conductor. Now disconnect the female portion of the connector attached to the receiver antenna lead and remove the small internal bushing and spring.

To assemble, slip the bared conductor through the female portion of the connector and then through the spring and bushing, making certain that the insulation enters the end of the connector. Bend over and spread the strands of the conductor against the forward end of the bushing and then force one of the eyelets (packed in small envelope in outfit package) into the bushing to hold the conductor in position. Cut off the ends of the conductor strands approximately ⅛ inch beyond the edge of the eyelet and bend the strands over toward the center of the eyelet. The assembly may be now attached to the receiver portion of the connector and the shield covering on the lead-in wire pushed forward to cover the adjacent end of the female portion. Finally, bond the shield to the connector by means of the small clamp furnished. **No soldering operations are required.**

NOTE—An unshielded lead-in wire (as in the case of the interior-type antenna) may be attached to the antenna connector as described above except that all references to the shield braid and loom may be neglected.

Power Supply to Receiver—The power input lead (*black* wire with fuse receptacle and terminal, extending from the receiver cable) must be connected electrically to the ungrounded side of the car storage battery. This connection preferably may be made at the battery terminal of the ammeter (usually the terminal with only one lead attached—consult wiring diagram in instruction book for automobile) and any slack length remaining should be taped securely behind the instrument panel.

Suppression of Ignition Interference

1. Disconnect all wires from the spark plugs. Fasten one spark plug suppressor to the top of each plug and re-attach the wires to the free ends of the suppressors. These suppressors may be mounted either in line with or at right angles to the plugs (as shown in Figure 1) in order to avoid interference with metallic parts grounded to the engine or frame.

2. If the distributor is of the plug-in type, disconnect the center wire from the head. Plug the distributor suppressor into the distributor head and insert the wire in the free end of the suppressor.

NOTE—For cap-type distributors, exchange the distributor suppressor at your dealer's for one of a special type. Cut the wire leading from the distributor to the coil and screw the suppressor into the end attached to the distributor. Screw the other end of the wire (leading to the coil) into the opposite end of the suppressor.

3. Clamp the generator by-pass capacitor against the generator frame. The screw holding the cut-out ordinarily may be utilized for securing this unit. Connect the capacitor lead to the terminal on the generator side of the cut-out switch. (In some cases, interference will be reduced by connecting the capacitor lead to the opposite side of the cut-out. The most suitable position for this lead must be determined by trial.)

4. The other by-pass capacitor must be connected between the battery terminal of the ammeter and any convenient screw on the instrument panel. In certain cases, interference will be reduced still further by connecting an additional capacitor (obtainable from your dealer) between the battery side of the ignition coil and the car frame.

PART II—OPERATION

The instrument should be operated as follows:

1. Insert the key in the lock on the remote control unit and turn it clockwise to the extremity of its rotation.

NOTE—This key serves to operate both the power switch and the volume control. A slight rotation clockwise will turn the power "on" and the remainder of the range permits adjustment of volume. The dial scale should become illuminated when the power is "on."

2. Rotate the Station Selector knob in either direction until a desirable station program is heard.

NOTE—The dial scale is calibrated in channels to aid in station identification. Add one cipher to the scale marking to obtain the actual frequency in kilocycles.

3. After receiving a signal, turn the Volume Control counter-clockwise until the volume is reduced to a low level. Now, readjust the Station Selector to the position midway between the points where

the quality becomes poor or the signal disappears. **This operation insures the best quality of reproduction.**

4. Finally, advance the Volume Control (clockwise) until the desired level is obtained. Except on weak signals, the automatic volume control will maintain the volume substantially at the latter level, thereby precluding further manual adjustments. (Fading of the signal may be experienced in extreme cases, as when passing under bridges or other metallic structures, since such structures almost completely shield the antenna.)

5. Set the Tone Range Switch (located on the front of the receiver unit) for the preferred tone quality. This switch has two positions. In the counter-clockwise position, high-frequency (treble) response and static interference (when present) are decreased.

6. When through operating, turn the key to the "off" position, counter-clockwise. The instrument is then locked by removing the key.

RCA-VICTOR, Inc.

SERVICE DATA

Type and Number of Radiotrons Used............1 RCA-89,
    1 RCA-78, 1 RCA-6A7, 1 RCA-6B7—Total, 4
Total Battery Current.................................5.5 Amperes
Undistorted Output....................................2.0 Watts
Loudspeaker Field Current.............................1.35 Amperes
Maximum Output D. C. Voltage from Rectifier...250 Volts
Total Plate Current...................................53 M. A.

This four tube Superheterodyne Automobile Receiver is of compact construction and gives excellent performance. Features such as unit construction (one unit contains the receiver, plate supply unit and loudspeaker), ease of installation, freedom from ignition noise and excellent sensitivity, selectivity and tone quality characterize this instrument.

Plate Supply Unit

This receiver uses a vibrator type Inverter and rectifier that provides a source of direct current voltage for use as plate and grid supply for all Radiotrons. *This unit is accurately adjusted at the factory and service adjustments should not be attempted.* Any difficulties with this unit should be referred to the nearest Distributor handling these instruments who has instructions for servicing this item.

Line-up Capacitor Adjustments

The three R. F. line-up capacitors and two I. F. tuning capacitors are accessible and may require adjustments. The R. F. adjustments are made at 1400 K. C. and the I. F. adjustments at 175 K. C. The R. F. adjustments can be made with the receiver in its case, access to the adjusting screws being obtained through a slot in the bottom of the case. For the I. F. adjustments, however, it is necessary to remove the rear cover in order to couple the oscillator to the first detector. The following procedure should be used for either adjustments:

R. F. Adjustment

The three R. F. line-up capacitors are adjusted at 1400 K. C. Proceed as follows:

(a) A fairly accurate adjustment can be made by using the ear for an indicating device, thus eliminating the need of an output meter and the necessity of removing the rear cover to connect it.

(b) Procure a modulated oscillator giving a signal at 1400 K. C. and a non-metallic screw driver.

(c) Couple the output of the oscillator from antenna to ground, set the dial at 140, and the oscillator at 1400 K. C.

(d) Place the oscillator and receiver in operation and adjust the oscillator output so that a weak signal is obtained in the loudspeaker when the volume control is at its maximum position.

(e) Then adjust the three line-up capacitors until maximum sound in the speaker is obtained. Readjust these capacitors a second time as there is a slight interlocking of adjustments.

For a more accurate adjustment, the use of an output meter is recommended. However, this will require the removal of the rear cover in order to connect the output meter across the cone coil. Also the bottom and Radiotron side of the chassis must be shielded together with the transformer so that vibrator noise will not be obtained, due to the removal of the case shielding.

I. F. Adjustments

In order to make the I. F. adjustments, it is necessary to remove the rear cover, due to the fact that the external oscillator must be connected between the control grid of the first detector and ground. Proceed as follows:

(a) Procure a modulated oscillator giving a signal at 175 K. C., a non-metallic screw driver and an output meter.

(b) Remove the receiver from its case, shield the transformer and Radiotrons as described under R. F. adjustments, place the receiver in operation and connect the oscillator output between the first detector grid and ground. Connect the output meter across the voice coil of the loudspeaker. Then connect the antenna lead to ground and adjust the tuning capacitor so that no signal except the I. F. oscillator is heard at maximum volume. With the volume control at maximum, reduce the external oscillator output until a small deflection is obtained. Unless this is done, the action of the A. V. C. will make it impossible to obtain correct adjustments.

(c) Each transformer has but one winding that is tuned by means of an adjustable capacitor, the other windings being untuned. The capacitors should be adjusted for maximum output.

At the time I. F. adjustments are made it is good practice to follow this adjustment with the R. F. adjustments, due to the interlocking that always occurs. The reverse of this, however, is not always true.

Practical Hints on Installation

The following suggestions may prove useful when making installations on the particular cars mentioned.

Chevrolet 1933—Mount chassis on left side, end against car bulkhead and use short flexible shaft. Use both capacitors, one on the ammeter and one on the generator. Use all suppressors. Place a copper screen under the toe board on right side, 10" x 10", to prevent the body from radiating ignition interference which may be picked up by the antenna. This screen must be grounded.

Plymouth 1933—Mount chassis on left side, back against car bulkhead and use 33⅞" flexible shaft. Use both capacitors, one on the ammeter and one on the generator. Use all suppressors.

Ford V-8 1932—Mount chassis on left side, end against car frame and use short flexible shaft. Use one capacitor, connected to the generator. Install eight spark plug type suppressors only, no distributor suppressor being necessary.

The majority of cars will be found to be entirely free from ignition noise when the standard equipment is used. Usually mounting the chassis on the right side of the bulkhead will be found most desirable, although if a heater is used, the left side will be preferable.

RADIOTRON SOCKET VOLTAGES

6.3 Volt Battery

Radiotron No.		Cathode to Ground	Cathode to Screen Grid Volts	Cathode to Plate Volts	Plate Current M. A.	Heater Volts
RCA-78 R. F.		3.7	92	253	7.0	6.06
RCA-6A7	First Detector	3.7	92	253	12.0	6.06
	Oscillator	0	—	253	Total	
RCA-6B7 Second Detector		3.2	92	236	6.0	6.06
RCA-89 Power		26.5	230	217	27.5	6.06

RCA-VICTOR, Inc.

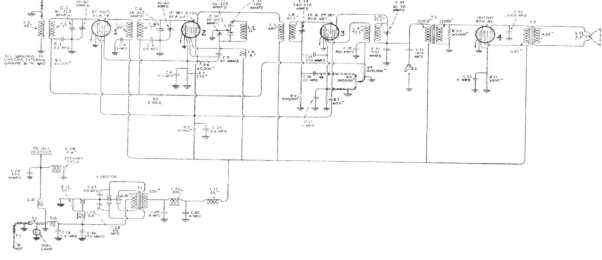

Figure A—Schematic Diagram

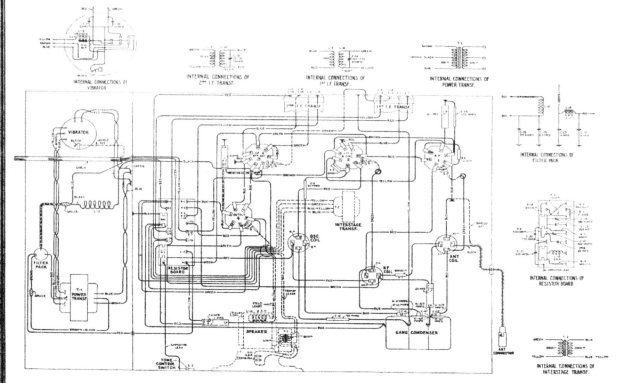

Figure B—Wiring Diagram

RCA-VICTOR, Inc.

SERVICE DATA FOR VIBRATOR UNIT

The vibrator unit used in this receiver is of excellent design and sturdy construction. It functions as a combined A. C. generator and mechanical rectifier. Referring to Figure C, it will be noted that the primary and secondary of the transformer are center tapped. By connecting the outside of each winding to the contacts of the vibrator and using the arms and center taps of the windings as sources of input and output voltage, a combined generating and rectifying action is obtained.

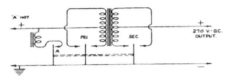

Figure C—Schematic of Vibrator Unit

When the switch is turned "on" the vibrator makes and breaks contact at point "A." This constitutes the driving action of the unit, and is in no way connected with the other circuits. The primary vibrator functions to connect the input low voltage current first across one-half and then across the other half of the primary of the transformer. This results in a pulsating direct current applied to the primary in an alternating direction. The result is an A. C. voltage emanating from the secondary of the transformer; as the transformer has a step-up ratio the A. C. secondary voltage is considerably greater than the primary. The secondary vibrator functions in a similar manner as that on the primary side, so that by reversing the alternations applied to the load, a pulsating D. C. is obtained. After filtering, this is used as plate and grid supply to all Radiotrons.

(1) Spring and Contact Adjustment Limits.

Proper adjustments of the various contacts are made in the following order and manner:

1. With 8 and 10, Figure D, firmly held against their respective stops and with 3 and 5 in contact with 8 and 10 respectively, the air gap between 1, 6 and 2, 7 shall be 0.015" plus or minus 0.005". On no particular unit however, shall the differences between the two air gaps exceed 0.005".

2. Adjust the buzzer screw, 11, Figure D, so that when the position of the armature is such that 1 and 2 are just making contact with 6 and 7 respectively, the contact between 4 and 9 shall just be breaking.

(2) Adjustment for the Reduction of Sparking.

If any pair of contacts show excessive sparking, the following procedure will in general reduce the sparking to a minimum.

For example, consider the case where excessive sparking is occurring between 6 and 1. Sparking will be reduced to a minimum by bending the armature spring on that side (secondary side) away from 6 and toward 8. (See Figure D.) If the bend is too small, only a small change will be noted. However, if an excessive bend is made, the sparking will be transferred from 6, 1 to 8, 3.

The same method may be applied to any pair of contacts. Usually only a slight bend will be necessary. Although after bending, no change in the position of the armature contacts may be noted, a sufficient change in the initial force requirements will have been made to reduce sparking.

(3) Output Voltage.

When connected to a 6 volt primary source, the output voltage across a 5,000 ohm resistor (connected in place of the receiver load at the output of the filter), must be 240 volts or greater.

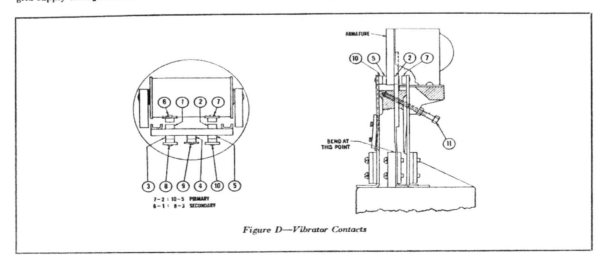

Figure D—Vibrator Contacts

SENTINEL RADIO CORP.

SENTINEL MODEL 561

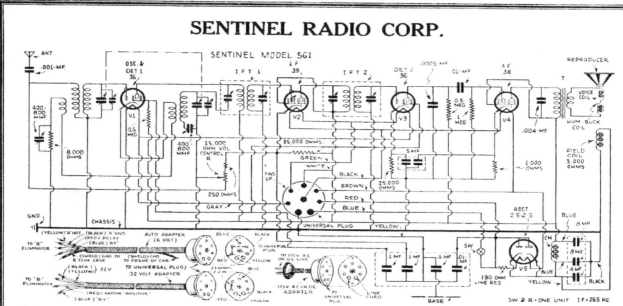

SENTINEL "A.C.-D.C." MODEL 561

PERHAPS the most interesting feature of this ultra-midget 5 tube set, illustrated, is its adaptability to the current supply, whether the available power is a 110 volt, A.C. or D.C. light line, a 32 volt farm lighting system, or the 6 volt storage battery in an automobile.

Included with the set are only the tubes, dynamic reproducer integral with the set, and a 25-foot reel of antenna wire. However, the following

accessories are available: a special Sentinel 125 volt, 60 milliampere interrupter-type "B" unit, car antenna, 6-volt adapter cable and plug, spark plug and generator suppressors, generator filter condenser, and mounting bracket assembly, web strap and buckle.

The receiver may be mounted in any convenient place in the automobile such as the robe rail in back of the front seat, between the dashboard and windshield pane, or on the underside of the dashboard head. Mounting accessories are available for this purpose.

In this receiver we find a superheterodyne circuit designed for A.C.-D.C. operation; the I.F. is 265 kc. Tubes used: One type 36 oscillator-first-detector; one 39, I.F. amplifier; one 36, second-detector; one 38, A.F. amplifier; and one 25Z5, rectifier. Total "B" requirement, about 60 milliamperes at 125 volts; a 90-volt block of "B" batteries may be used. The reproducer has a 3,000 ohm field.

Tube Type	Tube Position	Plate Volts	S.-G. Volts	C.-G. Volts
36	Osc.-Det. 1	112	25	2.5
39	I.F.	112	112	2.9
36	Det. 2	28*	25*	2.0
38	A.F.	108	112	1.5*
25Z5	Rect.	**		

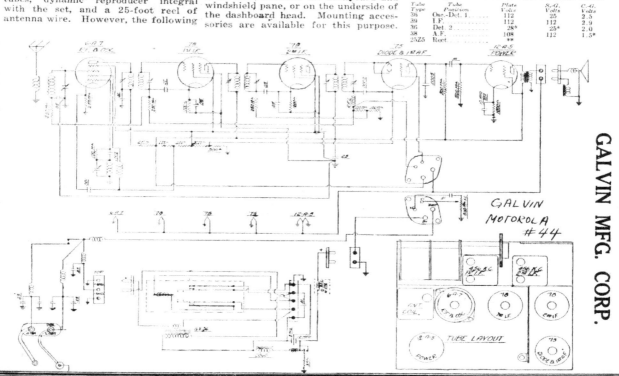

GALVIN MFG. CORP.

SPARTON MODEL 34

THE Sparton model 34 automotive receiver is a seven tube superheterodyne employing the Lafoy system of automatic volume control, pentode push-pull power output, and is equipped with finger tip remote-control tuning; the tuning control box is located on the steering column.

The receiver comes equipped with tubes but without accessories: the accessories include batteries and ignition suppressors. Either a wire mesh or a capacity-plate antenna may be used.

The mechanical arrangement of the set is such that it may be mounted either on the dash or under the floor-boards. The dynamic speaker, aside from the more conventional arrangement, may be mounted under the floor boards so that the grille is flush with the floor, like a heater.

Reference to the schematic circuit will show that a 39 is used as the first R.F.; a 36 as a combination detector-oscillator; a 39 as the I.F. amplifier tuned to 172.5 kc.; a type 70 second-detector and A.V.C. tube; a 37 in the first A.F. stage; and two type 38 pentodes in the output.

SPARKS - WITHINGTON, INC.

The battery equipment necessary, aside from the storage unit, is four 45 volt "B" batteries and a small "C" battery. The operating voltages and currents are as follows:

Tube Type	Plate Volts	Cont. Grid Voltage	Screen-Grid Voltage	Plate Curr.
V1	90	3.0	90	4.0
V2	120	15.0	90	2.0
V3	90	3.0	90	4.0
V4	180	...	...	1.0
V5	125	10.0	...	4.0
V6	180	19.5	180	8.0-10
V7	180	19.5	180	8.0-10

The above data are valid only when the condition of the batteries is good and the volume control is set to maximum with no signal.

A photograph of the equipment and a pictorial view of the receiver showing the location of all parts are appended.

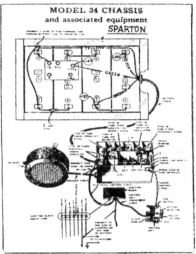

MODEL 34 CHASSIS
and associated equipment
SPARTON

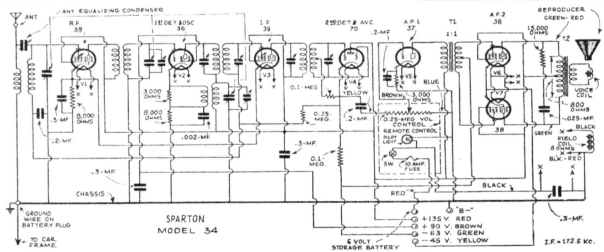

SPARTON MODEL 34

VOLTAGE ANALYSIS

Condition of "A" Battery—Good   **Position of Volume Control—Full with Antenna Disconnected**

Tube	Location	Filament Heater or	Plate	Control Grid —	Screen Grid +	Plate Current M. A.
'39	R. F. Stage	6.3	195	— 3.5	100	4.2
'36	1st Det.-Osc.	6.3	195	—12.	100	1.5
'39	I. F. Stage	6.3	195	— 3.5	100	4.2
85	2nd Det.-AVC	6.3	30	——	——	1.5
41	Power Stage	6.3	195	—15.	195	16.
84	Power Rec'fi'r	6.3	220	——	——	20 per plate

SPARKS - WITHINGTON, INC.

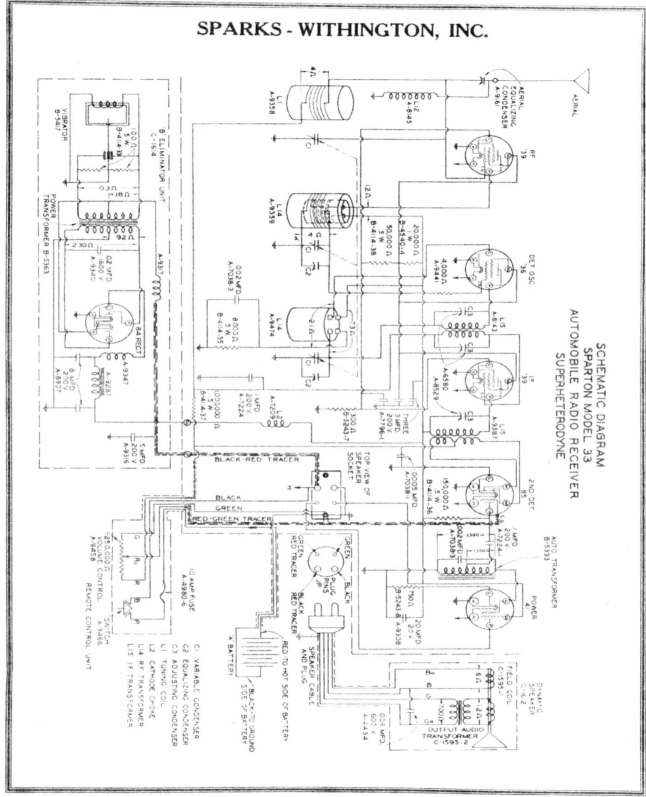

SCHEMATIC DIAGRAM
SPARTON MODEL 33
AUTOMOBILE RADIO RECEIVER
SUPERHETERODYNE

STEWART RADIO & TELEVISION CO.

STEWART "ARISTOCRAT"

THE Stewart "Aristocrat" is similar to the Stewart "Companion," here described, in many respects: the mechanical considerations are the same; but the electrical characteristics are different.

The sensitivity of this model is, also, 1 microvolt, measured by R.M.A. standards; the tubes used are as follows: a 39 as an R.F. amplifier; a 36 as a combination detector-oscillator; a 39 as an I.F. amplifier; an 85 as a combination second detector and A.V.C.; a 41, class A audio driver; a 79, class B power output tube; and an 84 rectifier. The "A" battery consumption of the set is 5.5 amperes at low signal levels, and 6 amperes at maximum power output.

The components of the "Aristocrat" and the "Companion" are completely interchangeable with the exception of the audio system and the "B" supply transformers.

STEWART "COMPANION"

THE Stewart Companion is a six-tube receiver of the superheterodyne type. It is 9⅜ inches wide, 6⅝ inches high, and 5⅞ inches deep. It is equipped with a remote tuning-control box, a combination "B" unit and speaker, and a receiver. The accessories are part of the car: they consist of all suppressors, condensers, bolts, washers, etc. The more conventional antennas may be used.

The receiver is designed to mount on the bulkhead, on either the engine or driver's side. The combination speaker and "B" unit mounts with a single bolt. The speaker should be so mounted that it may be adjusted, by means of the provisions provided, to suit the acoustic characteristics of the car, depending upon whether the car is crowded or not. Stewart Radio & Television Co. calls this feature "floating control."

As previously stated, the "Companion" is a six-tube superheterodyne using one 39 as an R.F. amplifier; a 36 as a composite detector-oscillator; a 39 as an I.F. amplifier; an 85 as a combination second-detector and A.V.C.; a 41 output pentode; and an 84 rectifier. The I.F. is 175 kc. The total current consumption of the set is five amperes at 6.3 volts; the sensitivity is 1 microvolt per meter, maximum, measured by R.M.A. standards; and the power output is 1.5 watts.

The receiver is equipped with manual quiet A.V.C., which the manufacturer calls "QAVC."

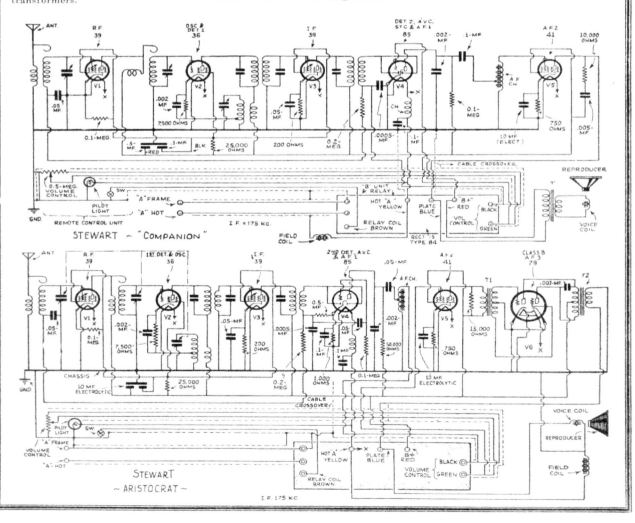

STEWART ~ "COMPANION"

STEWART ~ ARISTOCRAT ~

UNITED AMERICAN BOSCH CORP.

BOSCH 80 & 84

Frame

Cable Connector Plug

Bosch Motor Car Receiver

Control Unit

To "A" Supply

High (Red)

Gnd. (Black)

To "B" Supply

Blue   Brown   Green

R-1—Volume Control 18,000 ohms
R-2—1st RF Bias Resistor 500 ohms
R-3—Detector Bias Resistor 25,000 ohms
R-4—Detector Screen Resistor 500,000 ohms
R-5—Detector Plate Resistor 500,000 ohms

C-8—Detector Cathode Condenser .5mf.
C-9—Detector Screen Condenser .5mf.
C-10—Detector Plate Condenser .0001mf.
C-11—Detector Plate Condenser .0001mf.
C-12—Coupling Condenser .002mf.
C-13—Output Condenser 1.mf.
C-14—Filament By-pass Condenser
C-15—1st RF Alignment Condenser
C-16—3rd RF Alignment Condenser
C-17—Det. Alignment Condenser
C-18—Speaker Condenser

R-6—Audio Grid Resistor 2 meg.
R-7—Series Grid Resistor 250,000 ohms
R-8—Filament Resistor 1.3 ohms
R-9—Filament Resistor 1.1 ohms
R-10—Audio Bias Resistor 900 ohms
C-1—1st RF Tuning Condenser
C-2—2nd RF Tuning Condenser
C-3—3rd RF Tuning Condenser
C-5—Screen By-pass Condenser .5mf.
C-6—Cathode By-pass Condenser .5mf.
C-7—Plate By-pass Condenser 1.mf.

UNITED AMERICAN BOSCH CORP.

INSTALLATION INSTRUCTIONS

Car Radio Receivers should be installed by persons experienced in this particular line whenever possible. However, a brief set of instructions covering the installations of receivers is given below for the benefit of the purchaser of an automobile radio who might desire to install his own set or one who would like to familiarize himself with the generalities of this procedure.

The compact radio chassis, electro-dynamic speaker and "B" eliminator are in a single housing and mounted on the bulkhead of the car -- which is the partition between the engine and the interior of the car.

A suitable location for the set and control head should first be determined. This location should be such that the set mounted on the bulkhead should have adequate clearance and the control head mounted on the instrument panel should be conveniently located and at the same time not interfer with the operation of any of the controls of the automobile. The relative position of the set and control head should be such that the mechanical shafts, which inter-connect these two units, will not be bent at sharp angles.

The back plate of the receiver housing is detachable by means of two thumb screws, marked mounting thumb screws (Fig. 2). This back plate is also used in the mounting plate for the receiver and is attached to the bulkhead by means of three carriage bolts with suitable washers and spacers to accomodate the requirements of a particulr installation. The back plate, when removed, may be used as a template for locating the three holes in the bulkhead through which the mounting bolts are to pass. Care should be taken in locating these holes that the bolts upon passing through the bulkhead do not interfere with wires, tubing or other devices on the engine side of the bulkhead.

The mounting plate location having been established, The mounting plate should be attached to the bulkhead by means of the carriage bolt, spacers, washers and nuts provided for the purpose. Care should be taken to have a large washer located next to the interior surface of the bulkhead and that suitable spacers are provided between the interior and exterior surfaces of the bulkhead to prevent these two surfaces from being collapsed by the tightening of the screws.

Before the set is actually hung on the mounting plate, the car should be checked to determine which terminal of the battery is grounded. If the positive terminal of the car battery is grounded, connect the red wire on power pack terminal strip back of set to terminal #1 and the other wire to terminal #2. If the negative side of the car battery is grounded connect the red wire to terminal #2 and the other wire to terminal #1.

UNITED AMERICAN BOSCH CORP.

The actual attachment of the receiver to the plate is accomplished by placing the hook (attached to the upper rear corner of the housing) against the mounting plate and between the flanges of the latter. While still holding the receiver thus against the plate, the receiver is pushed upward until the hook catches over the plate; after being sure that the hook has caught all along the plate, the fingers should be run along the sides of the plate to insure that the receiver is between the flanges of the plate. The two thumb screws should then be installed beneath the receiver to secure it to the plate.

The connections to the car battery should be made at this period of the installation. The battery cable consists of two shielded wires one of which is provided with a fuse housing near the terminal. The RED battery wire should ALWAYS be connected to the "hot" terminal of the car battery and the BLACK terminal should ALWAYS be connected to the "grounded" terminal of the car battery. This is true regardless of whether the positive or negative terminal of the car battery is grounded. It is necessary, however, that the power pack connections be made as described above to establish proper polarity.

The battery cable, when properly routed for installation, should be clamped to the car at suitable points.

The shielded antenna cable, which is attached to the set, should be concealed and the free end of the conductor connected to the down lead from the roof antenna of the car. This joint should be soldered and securely taped to insulate the antenna against accidental grounding The connection should be made as high up in the windshield post as practicable, or as close to its lower opening as is possible. The shielding around the end should be grounded near its open end to the car frame.

The receiver should now be turned on by turning the volume control shaft in the set. The shaft end, referred to, is near the center of the front edge of the housing. Tune in a station at as high frequency (approximately 1400 kilocycles) as possible. The rotor of the tuning condenser in such a case will be well disengaged from between the stator plates. Ascertain the frequency, set the scale of the control head at that frequency and insert the flexible tuning shaft into the condenser pinion shaft end. Tighten the set screw on the pinion shaft end and then the one on the sleeve on the outside left end of the housing. This latter screw holds the shaft casing securely. If this adjustment has been made carefully, the calibration of the dial should follow closely the stations as tuned. Insert the end of the other flexible shaft into the volume control shaft coupling and tighten the set screw on the volume control shaft coupling.

The current drain from your car battery will be increased about 5 amperes while your radio is in use. It might be advisable, therefore to advance the third brush on your generator slightly to take care of this. This depends upon the conditions under which you operate your set.

UNITED AMERICAN BOSCH CORP.

OPERATION

As stated before, the compact radio chassis, "B" eliminator and electro-dynamic speaker, contained in a single housing, are mounted on the bulkhead of the car and a control unit is mounted on the instrument panel. (These components are shown diagramically in Fig. 2.) The set is turned on by operation of the left-hand knob on the control head. This knob also serves to regulate the volume of output.

The station selector is controlled by means of the right-hand knob on the instrument board control unit.

In tuning in station, the procedure should be as follows: First, turn the volume control knob to the extreme clockwise position of rotation and wait for the set to warm up as manifested by a slight crackling sound in the speaker. Then turn the station selector until the desired station is heard. After this, the volume can be regulated to the desired level by suitable counter-clockwise rotation of the volume control knob. The station selector knob should then be readjusted slightly so as to tune the station exactly to the most pleasing quality of output.

The numbers, which appear in the window of the control head in accordance with the rotation of the selector knob, indicate the frequency to which the set is tuned. These divisions are marked in hundreds of kilocycles so that the number 7 represents 700 kilocycles. The heavy line is used to indicate each of the hundred kilocycle points on the dial. The line of medium thickness is used to indicate the 50 kilocycles points, and a fine line to indicate the 25 kilocycle points occuring between the hundred kilocycle divisions.

The set may be locked by turning the set off and pulling out the volume control knob from the control unit. This knob should thereafter be used as a key.

ANTENNA

The Model 150 automobile receiver is designed for use in connection with the roof type of antenna, but if no such antenna is available, a capacitor plate can be obtained from the dealer, from whom you bought your set. This capacitor plate is mounted under the running board of your car. The use of this capacitor plate will eliminate the necessity and expense of installing an antenna in the roof of the car. However, a roof antenna will give better results and should be used wherever possible. Practically all of the 1933 closed car models are equipped with antennas in the roof, as are a great many of the 1932 models of closed cars. These roof antennas should be used whenever available.

UNITED AMERICAN BOSCH CORP.

SERVICING

The servicing of the Model 150 Car Radio should be done by a competent expert in this field. The American Bosch dealer and service station is equipped to provide you with information relative to service parts. He is also equipped to make minor repairs on the set and should be consulted wherever possible, in case of trouble. Troubles due to worn out or defective tubes or worn out or defective vibrator rectifiers may be corrected by withdrawal of these components from the set and substitution of similar components.

To remove the tubes, it is necessary to remove the five thumb nuts which hold the front of the set in place, and lay the front cover, with speaker attached, on the floor-boards of the car. The three tubes enclosed by the shield along side of the tuning condenser are then removed by placing a wire noose over the tube and pulling out. The other tubes can be pulled out directly with the fingers.

To remove the vibrator rectifier, the set must first be removed from the bulkhead and laid on the floor-boards. Then, with the front cover removed from the set, the vibrator rectifier can be ejected from its socket by pressing on the corners of the vibrator rectifier which project beyond the socket in the back of the set.

AMERICAN BOSCH VIBRO-POWER CAR RADIO

MODEL 150

Supplementary Service Instructions

The wiring diagram of the Model 150 American Bosch Car Radio shown on the other side of this page carries at the right hand side a list of electrical parts used. In ordering service replacements, please specify the "SA" number.

For the further information of the service man, the following data is given.

Total drain on the car battery: 5.1 amperes maximum.

Output: 1.7 watts.

Intermediate frequency: 175 kilocycles.

"B" voltage: 170 volts or more under set load.

UNITED AMERICAN BOSCH CORP.

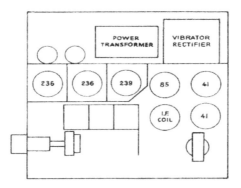

Fig. 1.

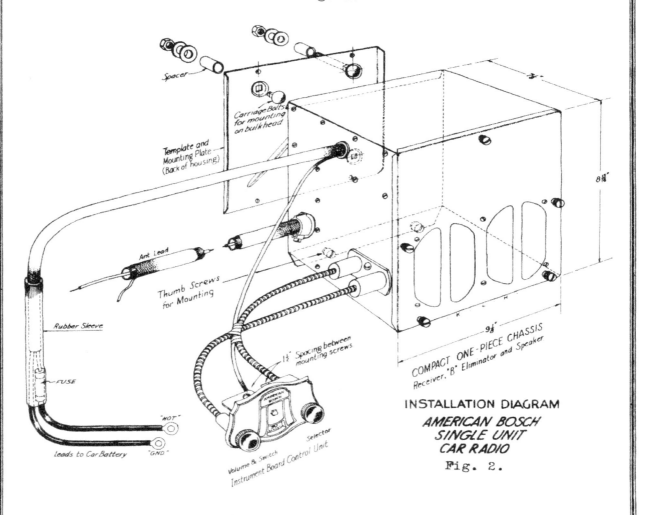

INSTALLATION DIAGRAM

*AMERICAN BOSCH
SINGLE UNIT
CAR RADIO*

Fig. 2.

UNITED AMERICAN BOSCH CORP.

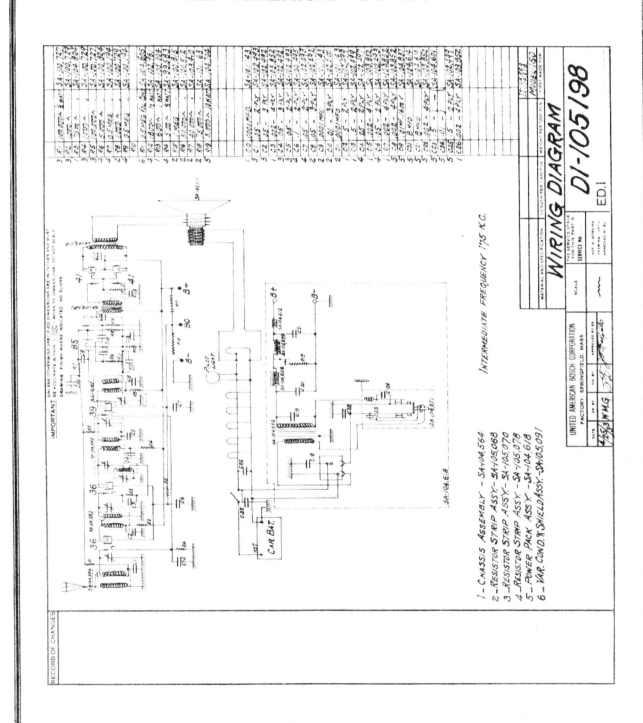

INTERMEDIATE FREQUENCY 175 K.C.

1 – CHASSIS ASSEMBLY – SA-104,564
2 – RESISTOR STRIP ASSY – SA-105,068
3 – RESISTOR STRIP ASSY – SA-105,070
4 – RESISTOR STRIP ASSY – SA-105,078
5 – POWER PACK ASSY – SA-104,618
6 – VAR. COND. & SHIELD ASSY – SA-105,091

WIRING DIAGRAM

DI-105198   ED.1

UNITED AMERICAN BOSCH CORPORATION
FACTORY: SPRINGFIELD, MASS.

UNITED MOTORS SERVICE
MODEL S-2035

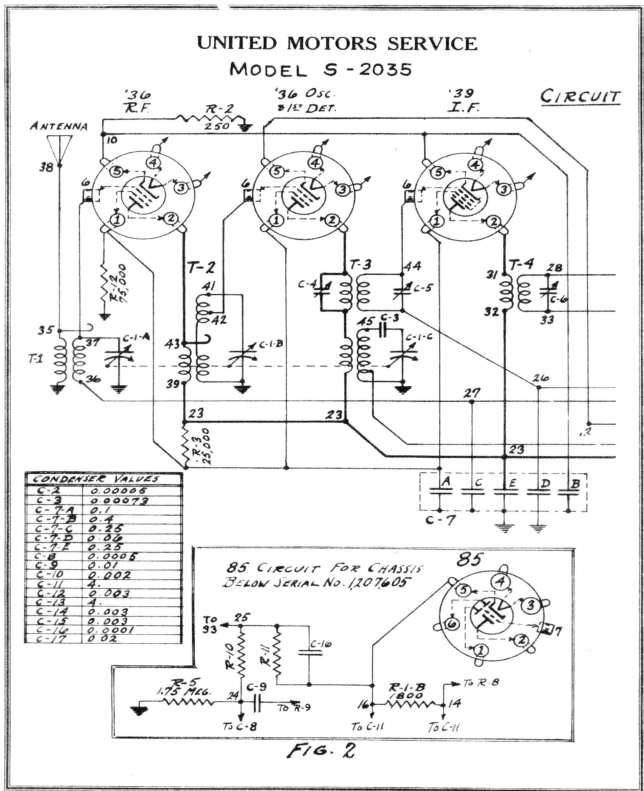

CONDENSER VALUES

C-2	0.00005
C-3	0.00073
C-7-A	0.1
C-7-B	0.4
C-7-C	0.25
C-7-D	0.06
C-7-E	0.25
C-8	0.0005
C-9	0.01
C-10	0.002
C-11	4.
C-12	0.003
C-13	4.
C-14	0.003
C-15	0.003
C-16	0.0001
C-17	0.02

FIG. 2

UNITED MOTORS SERVICE

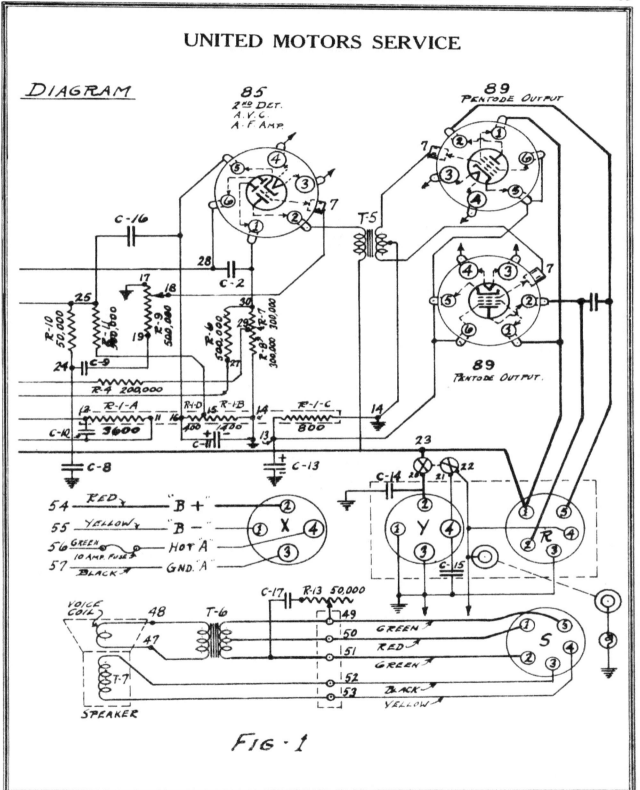

DIAGRAM

85
2ND DET.
A.V.C.
A.F. AMP.

89
PENTODE OUTPUT

89
PENTODE OUTPUT.

FIG·1

UNITED MOTORS SERVICE

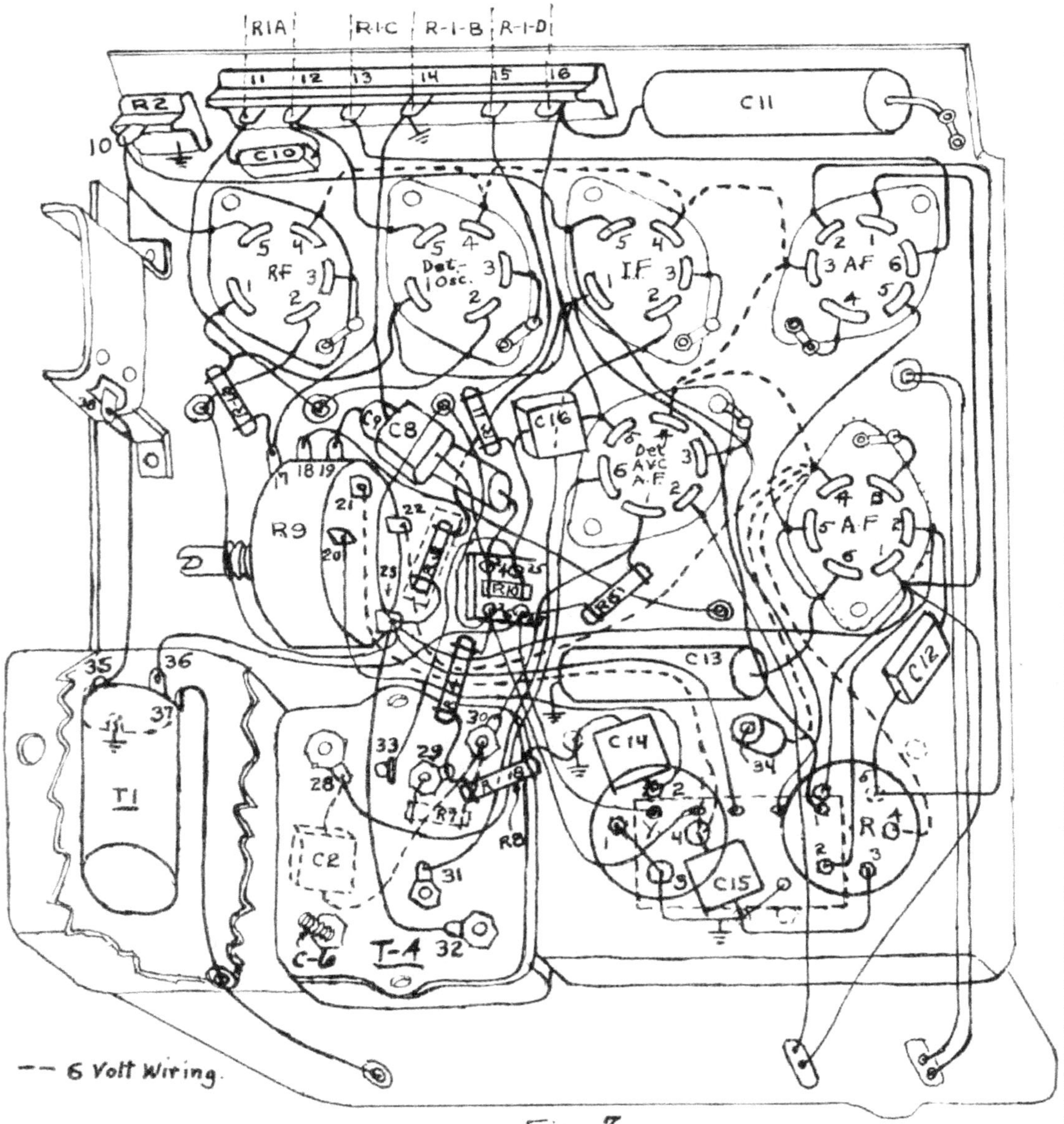

-- 6 Volt Wiring.

Fig - 3
Parts Location and Physical Wiring Diagram
Bottom View of Chassis

UNITED MOTORS SERVICE

Parts Location

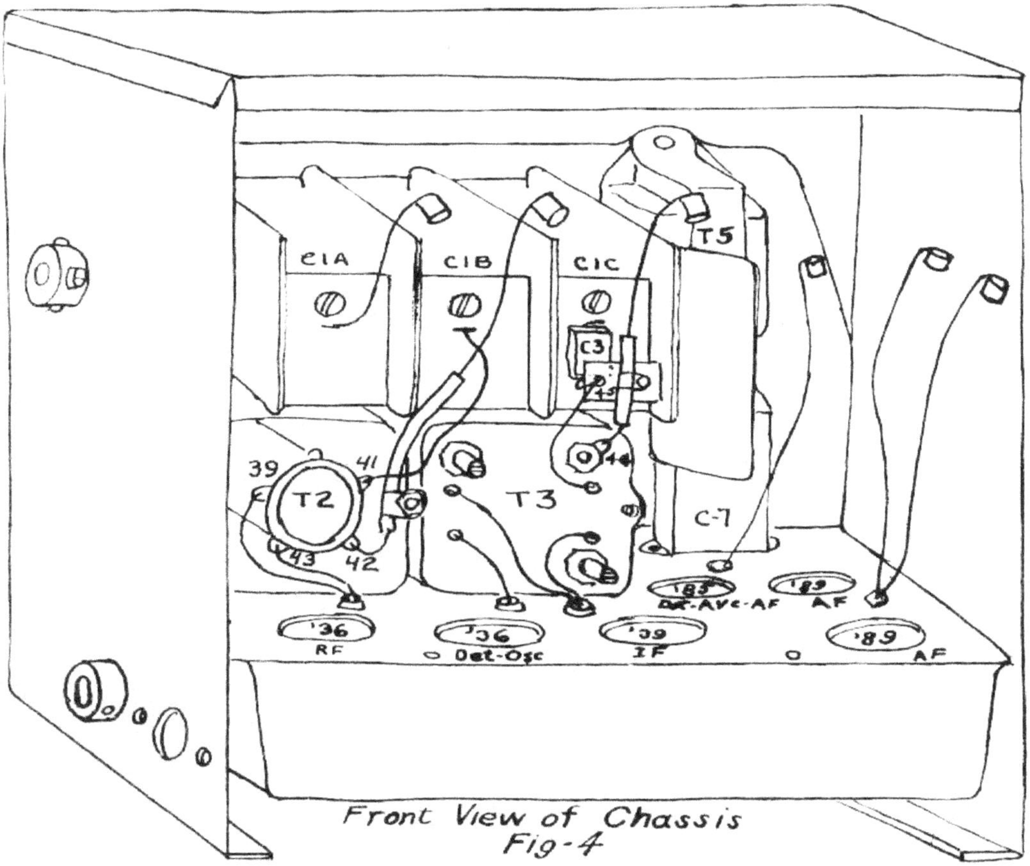

C1A   C1B   C1C   T5

C3

39 41 T2   T3   C-7   44

43 42

'36   '36   '59   '85   '89
RF   Det-Osc   IF   Det-Avc-AF   AF

'89
AF

Front View of Chassis
Fig-4

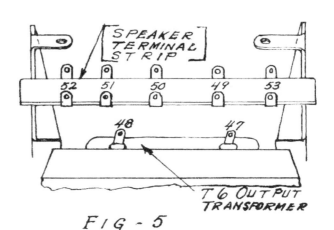

SPEAKER
TERMINAL
STRIP

52   51   50   49   53

48   47

T6 OUTPUT
TRANSFORMER

FIG - 5

UNITED MOTORS SERVICE

TROUBLE CHART (Complete Installation)--Cont'd.

A. NO RECEPTION (Selector dial not illuminated

  1. Fuse burned out in "A" battery cable.

  2. Improper connection at power supply Connector Plug.

  3. Poor connection between the Hot "A" battery wire and
     starter terminal or between the Ground "A" battery wire
     and the frame of the car.

  4. Inoperative Switch. First make certain that the switch
     is being actuated by the flexible control drive connected
     to the volume control knob. Then test the switch.

B. NO RECEPTION (Selector dial illuminated)

  1. Poor antenna, poor antenna connections or antenna
     disconnected.

  2. Shorted antenna clip (on chassis).

  3. One or more sections of C-7 Condenser Block shorted.

  4. Shorted R-5 Resistor.

  5. One or more defective tubes.

  6. Open in any circuit supplying voltage to tube sockets.
     (Take analyzer readings, see page 9.)

  7. Defective "B" Supply--Refer to "B" Supply Service Bulletin.

C. WEAK RECEPTION

  1. Poor antenna or poor antenna connections.

  2. One or more defective tubes.

  3. Worn out or weak "B" Batteries. If Dynamotor is used,
     check output of same to insure correct voltage.

  4. Open speaker field.

  5. Open or shorted C-7-A, C-11, C-13 or C-16 condensers.

  6. Shorted C-2 condenser (Diode plates shorted).

  7. Shorted R-12 Resistor.

UNITED MOTORS SERVICE

TROUBLE CHART (Complete Installation) --Cont'd.

C. WEAK RECEPTION--Cont'd.

8. Short between the wiring and parts connecting to the diode plate of the type 85 tube socket and ground.

9. Trimmer condensers out of alignment. (See instructions for aligning condensers on page 10.)

D. POOR TONE QUALITY

1. Low "B" batteries. Replace any "B" batteries testing below 35 volts with new batteries. If Dynamotor is used, the output should be above 170 volts.

2. Low "A" (car storage) Battery. The filament voltage when measured with a set analyzer should never be below 5.5 volts

3. One or more defective tubes.

4. Defective speaker.

5. Station not properly tuned in.

6. Defective chassis (Remove from car and make complete test).

E. BLOCKING (SPUTTERING)

1. Open C-7-C or C-7-D section of condenser block. (This condition will manifest itself by the R. F. or I. F. tube drawing grid current and developing high bias.)

2. Shorted R-3 resistor.

3. Open C-10 Condenser.

4. Abnormal 236 Detector-Oscillator tube. Install new tube.

F. OSCILLATION WITHOUT DECREASE IN SENSITIVITY

1. Open C-12 Condenser.

2. Open C-8 Condenser.

3. Shorted R-10 Resistor.

4. Grid lead coupling caused by the leads of one stage being moved into close proximity to those of another stage.

5. Coupling between antenna and speaker connections. Separate antenna lead and speaker cable as far as practicable.

UNITED MOTORS SERVICE

TROUBLE CHART (Complete Installation) --Cont'd.

G. OSCILLATION WITH DECREASE IN SENSITIVITY

    1. Open section of C-7 Bypass condenser. Note: When section C-7-C is open, there is also a tendency to block, while section C-7-D opening will give a tendency toward "motor boating."

    2. Open R-12 resistor (Causing high screen voltages).

TESTING ACCESSORIES (Antenna, Tubes, Batteries, etc.)

A. Antenna. Substitute a piece of insulated wire, approximately 6 or 8 feet long and strung out through the door of the car, for the regular antenna. Then, if the set operates normally, the regular antenna is at fault and should be repaired.

B. Tubes. Remove one at a time and test carefully, or substitute known good tubes.

C. Batteries. Remove the battery cable plug from the chassis and with a direct current voltmeter, having a double range (0 to 8 and 0 to 200), make the following tests (see Fig.1):

TEST		Correct	Location of
Pos.	Neg.	Voltage	Trouble
X-4	X-3	6	"A" Battery
X-2	X-1	135	"B" Batteries
X-2	X-1	200	Dynamotor
X-2	X-1	200	Eliminator

D. Battery Cable and Connections. Make the tests outlined in "C" and again directly at the terminals of the batteries. Then if substantially the same readings are not obtained, either the connections are not made properly or the cable is defective.

E. Speaker. Remove the speaker cable plug from the socket located on the chassis and plug in (without mounting) a known good speaker. Then if the reception is normal, it is an indication that the speaker is defective and should be removed, tested and repaired.

F. Chassis. Having determined that all accessories are in good operating condition, that all battery connections are made properly and that the fuses are not burned out, it will then be necessary to remove the chassis from the car and connect it up on the test bench. Then make all tests listed under "Volt-Ohmmeter Test of Chassis" or "Testing with a Set Analyzer."

UNITED MOTORS SERVICE

TESTING WITH VOLT-OHMMETER

To simplify testing the chassis, all contacts used in making the tests are numbered--see Figs. 3 and 4. The same numbers will also be found in their respective positions in the Wiring Diagram, Fig. 1.

VOLTAGE TESTS:

The correct approximate voltage for each tube socket prong, when using a Model D-100 Dynamotor or B-101 Eliminator, is listed in the following chart. When testing with three 45 volt "B" Batteries, the voltages should be about 75% of those listed.

Touch the positive prong of the test leads to tube socket terminals, one at a time, and the negative prong to the frame of the chassis. IMPORTANT--ALL VOLTAGES UNDER 6 SHOULD BE MEASURED ON 6 VOLT SCALE.

VOLTAGE CHART

Tube	Screen Contact #1	Plate Contact #2	Heater Contact #3	Heater Contact #4	Cathode Contact #5	Supressor Grid Contact #6
236 RF	80	170	0	6	2.4	- -
236 OSC	80	170	0	6	5.5	- -
239 IF	80	170	0	6	2.4	- -
85 DET	* 0	160	0	6	10.5	0.15**
A-89 AF	170	170	6	0	20.5	20.5
B-89 AF	170	170	0	6	20.5	20.5
Speaker-Socket	170	170	0	6	0	- -

Also test from 85 Det. Contact #1 (Pos.) to Contact #5--should read 3 volts on 6 volt scale. Incorrect reading indicates defective R-1-B, R-1-D, R-7 or R-8 Resistor.

NOTE: After checking the above voltages, if any are incorrect-- refer to "Locating Troubles Isolated by Voltage Tests" to find part at fault.

* #1 Terminal on type 85 tube connects to A.V.C. plate
** #6 Terminal on type 85 tube connects to detector plate

UNITED MOTORS SERVICE

TESTING WITH VOLT-OHMMETER--Cont'd.

OHMMETER TESTS: It is necessary to make the following resistance measurements, with tubes removed, in addition to either the voltage tests or set analyzer tests, in order to make a complete check of the chassis.

Test From	To	Correct Resistance (In ohms)	Probable Location of Trouble If Incorrect Reading is Obtained
1. Ground	38	30	T1 Antenna Coil Pri.
2. 236 R.F. #6	27	6	T1 Antenna Coil Sec.
3. 236 OSC #6	41	4	T2 R. F. Coil Sec.
4. Ground	41	9	T2 R. F. Coil Sec.
5. 239 I.F. #6	26	50	T3 Osc. Coil
6. Ground	45	5	T3 Osc. Coil
7. Ground	11	1	T3 Osc. Coil
8. 85 Det. #6	25	50	T4 I. F. Transformer
9. Speaker Soc. #2	5	Open	C-12 Condenser
*10. #48	47	3	T6 Output Trans. Sec.
11. #24	25	50,000	R10 Resistor
12. Voice Coil Lead	47	0.5	Speaker Voice Coil
13. A-89 A.F. #7	GND	4,000	T5 Input Trans. Secondary
14. B-89 A.F. #7	GND	4,000	T5 Input Trans. Secondary
**15. 85 Det. #7	GND	zero to 500,000	R9 Volume Control
16. #26	29	200,000	R4 Resistor
17. #27	30	50,000	R6 Resistor

 * Unsolder and disconnect Voice Coil Lead from Terminal #48
** Rotate Volume Control

LOCATING TROUBLES ISOLATED BY VOLTAGE TESTS

(By means of Resistance Measurements)

The Volt-Ohmmeter test of the chassis, or the "Set Analyzer" method of testing, only isolates the source of trouble in some one stage of the complete circuit, for example, the 1st Detector Stage or possibly the I. F. Stage. Then the actual part at fault may be easily located by making the resistance measurements listed opposite the condition disclosed by the voltage tests recorded in the following chart:

UNITED MOTORS SERVICE

LOCATING TROUBLES ISOLATED BY VOLTAGE TESTS--Cont'd.

(By Means of Resistance Measurements)

Description of Incorrect Voltage	Test From	To	Correct Reading (In Ohms)	Part or Parts Probably Causing Incorrect Voltage
A. No. Filament (A) Voltage at any Socket	1. Hot "A" Lead	X4	Zero	Fuse, or Green
	2. Ground	Y4	*Zero	Lead of "A" Cable Switch or Wiring
B. No. Plate (B) Voltage at any Socket	1. 54	X2	*Zero	"B" Cable
	2. 23	Y2	*Zero	Switch ("B" Sec.)
	3. Ground	Y2	100,000	C-14 Condenser
	4. Ground	23	100,000	C-7-E Condenser
C. "89" Sockets				
a. Plate Voltage	1. R2	R5	Open	C-12 Condenser
	2. S2	S5	900	T6 Transformer
	3. S2	S1	450	T6 Transformer
b. Screen Voltage	1. R1	A1	Zero	Wiring
	2. R1	B1	Zero	Wiring
c. Cathode Voltage	Ground	B5	800	C-13 Condenser or R-1-C Resistor
d. Supressor Grid Voltage	Ground	B5	800	C-13 Condenser or R-1-C Resistor
D. 85 Socket				
a. Plate Voltage	85 Det. #2	23	2600	T5 Transformer
b. A.V.C. Plate or Det. Plate Voltage	1. 85 Det. #6	25	5	T4 Transformer
	2. 24	25	50,000	R10 Resistor
	3. 85 Det. #5	25	500,000	R11 Resistor or C-16 Condenser
	4. 85 Det. #1	29	300,000	R-7 (Encl. in T4)
	5. 29	14	300,000	R-8 Resistor
	6. 15	16	400	R-1-D Resistor
	7. 15	14	1,400	R-1-B Resistor
**8. 14		16	1,800	R-1-B Resistor
	9. 85 Det. #1	28	1,100,000	C-2 Condenser
	10. Ground	26	500,000	C-7-D Condenser
	11. 24	19	Open	C-9 Condenser
	12. 24	GND	551,000	C-8 Condenser

* Switch--on
** Disconnect C-11 Electrolytic Condenser and Test separately

UNITED MOTORS SERVICE

LOCATING TROUBLES ISOLATED BY VOLTAGE TESTS--Cont'd.

Description of Incorrect Voltage	Test From	To	Correct Reading (In Ohms)	Part or Parts Probably Causing Incorrect Voltage
E. 39 I. F. Socket				
a. Plate Volts	I.F. #2	23	5	T-4 Transformer
b. Screen Volts 1.	I.F. #1	23	25,000	R-3 or C-7-A
2.	I.F. #1	GRD	75,000	R-12 or C-7-A
c. Cathode Volts	I.F. #5	GRD	250	R-2 or C-7-B
F. Osc. & 1st. Det.				
a. Plate Volts	Osc. #2	23	5	T-3 Coil
b. Screen Volts 1.	Osc. #1	23	25,000	R-3 or C-7-A
2.	Osc. #1	GRD	75,000	R-12 or C-7-A
c. Cathode Volts	Osc. #5	11	*3,600	R-1-A or C-10
G. 236 R. F. Socket				
a. Plate Volts	R.F. #2	23	5	T-2 Coil
b. Screen Volts 1.	R.F. #1	23	25,000	R-3 or C-7-A
2.	R.F. #1	GRD	75,000	R-12 or C-7-A
c. Cathode Volts	R.F. #5	GRD	250	R-2 or C-7-B
H. Speaker				
a. Weak	S3	S4	6	T-7 Speaker Field
b. Distorted 1.	S1	S5	900	T-6, C-17 or R-13
2.	46	49	50,000	R-13 Resistor

NOTE: It will be necessary to disconnect one lead of C-2, C-7
(All Sections), C-10, C-11, C-14, C-16, C-17 Condensers
in order to test them accurately

Refer to "Testing Electrolytic Filter Condensers" for
details on testing C-11 and C-13 condensers.

* R-1-A Resistor originally measured 4200 ohms. This was changed
to 3150 ohms at Serial No. 1207605, to 4000 ohms at Serial
No. 1207761 and finally to 3600 ohms at Serial No. 1222409.

UNITED MOTORS SERVICE

TESTING WITH A "SET ANALYZER"

The following chart shows the approximate readings that should be obtained with any one of the more reliable makes of radio set analyzers.

Readings obtained with set analyzers will vary with different makes of analyzers, with voltage variations and with different tubes. The readings shown in the table, page 12, therefore, are really only average values. Each service man should calibrate the chart of his service manual to match his set analyzer, using a radio set that is known to be operating properly.

Make a complete test, at each socket, with the set tuned between stations and the volume control on full, and list readings obtained in a similar form to the chart shown so that an easy comparison can be made in order to isolate the trouble in some one circuit.

A motor boating may sometimes develop when the Set Analyzer is being used, due to coupling set up in the analyzer cable. This motor boating may be eliminated by connecting a fixed condenser (about .1 M.F.) from the grid cap of the tube being tested to the frame of the chassis.

Then after a complete chart has been prepared for the set being serviced, compare that chart with the standard chart on the next page and note the stage of the circuit in which there is a difference in voltage. The information given under the subject "Locating Troubles Isolated by Voltage Tests" will make it possible to easily and quickly locate the part or parts in that stage causing the incorrect voltage.

The voltages shown were taken while using a model B-101 eliminator as the "B" supply. The same approximate voltages should be obtained when using a model D-100 Dynamotor. However, if three 45 volt "B" Batteries are used in place of the eliminator, all voltages will be approximately 75% of those listed.

UNITED MOTORS SERVICE

TESTING WITH A "SET ANALYZER"--Cont'd.

		Voltages			Milliamperes			
Tube	Position	Plate	Screen	*Cath-ode	Fila-ment	Plate	Screen	Grid Test
236	R. F.	170	80	2.4	6.0	2.5	.01 to .15	2.8
236	1st Det.	170	80	5.5	6.0	.8	.01 to .04	.8
239	I. F.	170	80	2.4	6.0	3.7	1.03	.4
85	(A.V.C.) (2d Det.) (A. F.)	160	--	4.5	6.0	4.	- - - - -	-
89	Power	170	170	20.	6.0	7.5	1.1	- -
89	Power	170	170	14.	6.0	7.5	1.1	- -

*NOTE: The Cathode voltage (20) listed for one of the type 89
tubes is the actual cathode voltage applied to the element of
the tube. However, the voltages listed for the remainder of
the tubes read bucking filament volts on normal cathode scale.

For example: The reading obtained for the other 89 tube should
be 20 volts minus 6 volts, or 14 volts.

If any readings obtained are very dissimilar to those listed on
your calibrated analyzer chart, it will be easy for the experienced
serviceman to determine the stage in which the trouble is located,
and by referring to the wiring diagram, Fig. 1, and the "Parts
Location" drawing, Fig's. 3 & 4, each part making up that circuit
may be tested individually until the one at fault is found.

TESTING ELECTROLYTIC FILTER CONDENSERS

There are two dry electrolytic condenser units, Part No. 1207616.
Both of these condensers are identical in construction, having a
capacity of 4. M.F. These units act as condensers only when the
anode is kept at a positive potential. Upon the application of
an excessive voltage or reversed polarity, the film which forms
the dielectric, breaks down and the condenser action ceases. However, no damage will result, as upon return to normal condition,
the film will rapidly build up and the unit will again function
as a condenser. In making continuity tests on the chassis this
fact should be kept in mind. For any tests that are across the
electrolytic condensers, the positive prod should be applied to
the point that is connected to the anode of the condenser. If
the anode is made negative, the condenser will pass current much
more readily than if it is made positive and the reading will be
different than the standard reading which should be obtained.

UNITED MOTORS SERVICE

TESTING ELECTROLYTIC FILTER CONDENSERS--Cont'd.

If there is any doubt about the condition of the electrolytic condensers, they should be checked for leakage current. This is done by applying a D.C. potential of 300 volts to the terminals of the condensers, the anode of course being made positive. After being on test five minutes, measure the leakage current with a milliammeter (which should be shunted for at least five minutes as at first there is a high current through the condenser which gradually drops away as the film forms). Any condensers reading 1. M.A. or less (per 4 M.F. Section) will operate satisfactorily.

In testing electrolytic condensers for capacity, ordinary methods involving the use of A.C. potentials cannot be used as the anode must always be kept positive. If there is any doubt about the capacity of an electrolytic condenser unit, it is recommended that a new one be secured and that the old one be checked by means of a comparison test.

In replacing the electrolytic condenser units, extreme care should be taken to wire them in with the correct polarity. Tag the leads when they are taken off the old condenser. The positive terminal of the condenser is identified by a positive symbol on the box. The positive lead in the chassis can be determined by referring to the circuit diagram.

TESTING ELECTRODYNAMIC SPEAKER

DESCRIPTION--The field of the speaker has a resistance of 6 ohms, and is connected in parallel with the filaments of the tubes. The output transformer, which is mounted inside the speaker case, is accurately designed to match the load impedance of the type 89 output tubes and the impedance of the voice coil.

Do not use any other type of speaker than the type supplied, as the speaker is especially designed for the receiver.

CONTINUITY TESTS--Continuity Tests on the primary of the output transformer and the field coil may be made at the terminal strip inside of the speaker case or at the speaker plug. (See page 13 for circuit.)

To test the secondary of the output transformer and the voice coil, it will be necessary to unsolder one of the voice coil leads at the terminal on the output transformer in order to test through each winding separately. Neither winding should show a ground to the frame of the speaker.

To test the tone control rheostat and condenser, disconnect the condenser from the rheostat and then test each unit separately.

UNITED MOTORS SERVICE

ALIGNING CONDENSERS

PEAKING ADJUSTABLE CONDENSERS

All of the adjustable condensers, commonly called trimmer condensers, are very accurately adjusted at the factory and will not need any further adjustment unless a coil or I. F. transformer is changed or the adjustments are tampered with in the field.

DO NOT attempt to change the setting of any of the trimmer condensers unless it is definitely known that adjustment is necessary, and an accurate test oscillator and a screw driver (with fibre handle) are available. Using a standard metal screw driver for this purpose will not give accurate adjustment.

Proceed as follows:

A. Disconnect the antenna lead-in from the chassis.

B. Ground the antenna terminal on the chassis to the frame of the chassis.

C. Set "test oscillator" to 262 kilocycles. Some oscillators are not equipped with a frequency of 262 K.C. but do have a frequency of 130 K.C. In this case, the second harmonic of 130 K.C., namely 260 K.C., may be used.

D. Connect the output leads of the test oscillator to the grid of the 1st Detector tube and to ground (frame of the chassis). Leave grid cap in place.

E. Connect an output meter in parallel with voice coil of the speaker (make connections at terminal strip in the speaker), or across the plates of the type 89 tubes with a 1 M.F. condenser in series.

F. Turn the tuning condenser rotor to minimum capacity (rotor plates out of stator places).

G. Adjust I. F. Trimmers in the following order, in each case leaving the trimmer set for maximum output as shown by the output meter:* (See note at bottom of page 15.)

> * C-4, Plate circuit of 1st Det.
> C-5, Grid circuit of I. F. Amp.
> C-6, Diode Input circuit.

   * See Fig. 3 for location of condensers.

UNITED MOTORS SERVICE

ALIGNING CONDENSERS--Cont'd.

H. Remove connection grounding the antenna (reverse of instructions under B).

I. Insert the Calibration Block, Part No. 1206418, between the center (2nd R. F.) condenser and the rear of the chassis as follows: Lay the block on the bench with the largest flat side down and the cut-out edge toward the operator. Pick up the block between the first and second fingers of the hand so that the side having the beveled and cut-out edges faces the knuckles of the hand, and the fingers are as close to the beveled corners as is possible. Insert the hand in the case over the center tuning condenser (condenser plates fully closed) and place the Block between the condenser bracket and the chassis back, with the largest face of the Block flat against the back of the chassis. The Block will fit quite tightly and the left side must rest against the shield between the 1st and 2nd R.F. condensers in order to clear the condenser wiper spring.

J. Attach the test oscillator to antenna terminal and ground (frame) of the chassis. (Ant. on test oscillator to Ant. on chassis and ground on test oscillator to frame of chassis.)

K. Set test oscillator at 1400 K.C.

L. Open tuning condenser until it stops against the Calibration Block.

M. Place Tube Shield in position around 236 Det.-Osc. tube. Adjust the trimmer condensers on the tuning condenser to maximum output, as measured by the output meter, in the following order:

```
C-1-C--Oscillator trimmer
C-1-B--2nd R. F. trimmer
C-1-A--1st R. F. trimmer
```

*NOTE: To insure sharp peaking of all trimmers, set the oscillator output below the point of start of A.V.C. action. Either set the output of the oscillator so that it is less than half the maximum output available, or use an 85 tube with the Diode A.V.C. plate prong removed (Prong #1).

CALIBRATING THE TUNING CONDENSERS

A. Remove the top of the chassis case.

UNITED MOTORS SERVICE

ALIGNING CONDENSERS--Cont'd.

B. Insert the calibration Block, Part No. 1206418, in between the chassis and the middle tuning condenser. The largest flat side must be against the back of the chassis and the Block must fit against the left side of the condenser shield. (See paragraph H, "Peaking Adjustable Condensers," for details.)

C. Insert the Calibration Dial drive shaft through the condenser drive ferrule into the drive pinion and turn the shaft until the condenser stops against the Block. Tighten one set screw in the drive ferrule to lock the condensers against the Block.

D. Slide the Calibration Dial, Part No. 1206421, on the drive shaft until it touches the ferrule.

E. Fasten the Indicator Strip, Part No. 12064 in place by mounting it under a thumb nut on the lid screw of the left front edge of the chassis case.

F. Turn the Calibration Dial until the 1400 line in the outside circle is located at the upper edge of the Indicator Strip; tighten the two set screws in the Dial hub.

G. Attach the test oscillator to the antenna terminal and ground (frame of chassis).

H. Set the oscillator at 1400 kilocycles.

I. Place the Tube Shield, Part No. 1206419, in position around the Det.-Osc. tube.

J. Peak the trimmers as indicated in M of the peaking instructions.

K. Release the set screw in the ferrule and tune the receiver to 1000 K. C. as indicated by the Calibration Dial and Indicator Strip.

L. Turn the oscillator to 1000 K.C.

M. Note the position of the trimmer on C-1-B condenser and then screw it down slightly. If the output as measured on the output meter increases, replace the trimmer in its original position. Now bend the outside split plates, which are partly in mesh with the stator plates, toward the stator plates until maximum output is obtained.

     If turning the trimmer screw down decreases the output, try unscrewing the screw. If a peak is obtained with the screw

UNITED MOTORS SERVICE

ALIGNING CONDENSERS--Cont'd.

out further than its original position, bend the plates away from the stator plates until maximum output is obtained.

If the peak occurs at the original position of the trimmer screw, do not bend the plates.

NOTE: Use a pair of duck-bill pliers for bending the plates and do not leave any sharp bends in the plates. Bend the plates so they form a smooth curve--do not have one plate bent in and the next plate bent out a quarter of an inch unless the plates are bent in the shape of an S.

N. Repeat "L" with C-1-A Condenser.

O. DO NOT attempt to bend the C-1-C oscillator condenser plates.

P. Repeat K, L, M, N and O at 750 and 600 K.C.

Q. Repeat M of "PEAKING ADJUSTABLE CONDENSERS." The Calibration Dial and shaft, Indicator Strip, Calibration Block, Tube Shield and antenna lead are grouped under part number 1207770. This kit simplifies the peaking and calibration of the Model 2035 United Motors radio receivers.

In order to accomplish the above peaking and calibration instructions without the calibration kit, it will be necessary to build a frame to hold the receiver case and control drive rigid with respect to each other. First log the dial in the control drive by tuning in a station of known frequency below 700 K.C. and adjust the drive so that the dial indicates that frequency. Now follow the above instruction for peaking and calibrating the condensers, using the control drive to determine 1400, 1000, 750 and 600 K.C.

EMERSON ELECTRIC MFG. CO.

EMERSON "B" UNIT

THE Emerson "B" power unit of a dynamotor type, having a sealed filter pack mounted on steel end-plates, measuring 6⅞x4⅝x7⅜ inches, has been made available through the Emerson Electric Mfg. Co., Inc. This unit may be used on sets which obtain their "C" bias from "B—" to ground or on sets which have the "B—" grounded. The machine is rated at 180 volts, 40 milliamperes, and draws 2½ amperes at 6.3 volts. This unit has two armature windings which rotate in a common field. The field coil, of course, is connected across the low-voltage winding.

A 90-volt tap is provided for receivers requiring this voltage. This tap is secured by connecting 15,000- and 20,000-ohm resistors in series across the output of the unit; the 90-volt tap connects to the junction of these two resistors. A 6 mf. condenser is used across the high voltage terminals, as shown in the schematic here. A 2 mf. condenser isconnected from "B—" to ground and a 1 mf. condenser from "A+" to "A—." The unit should be lubricated but once a year using a medium weight mineral oil, such as a light automobile engine oil .

Efficiency, output voltage, and current drain curves are also shown.

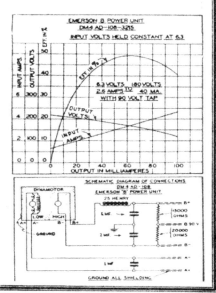

WELLS - GARDNER COMPANY
AUTOMOBILE RADIO
RECEIVER
062 SERIES
SUPERHETERODYNE

THERE are certain fundamentals about the installation of an auto set that should be studied and considered seriously before any attempt is made to install the receiver in the car. No attempt will be made in this manual to state definitely how the set should be installed in each make of car. Instead we will go through step by step the method of installing each unit of the set, giving the different methods that may be used in installing these units. *One of these methods will be applicable to any make of car.* The type of installation should be determined upon for the particular car on which the set is installed before starting the job. The different methods of installation with diagrams and complete installation data are given in other parts of the service manual, and these should be carefully checked over and *the best method of installing each unit determined before any attempt is made to install the receiver.* The chassis (unit containing tuning unit and tubes) may be installed in any one of a number of locations; steering column, on the dash underneath the cowl or on the dash under the hood. The "B" eliminator or "B" battery box can be mounted at any convenient position under the seat or under the car.

In mounting the "B" eliminator it is preferable to mount it at least three feet away from the radio chassis, under the hood, or in any one of the locations shown in the manual. It should be installed so that it will not interfere with the operation of any of the controls on the car, and not placed near the exhaust pipe. The chassis may be installed in any one of the locations shown in the manual, and in such a position that it will not interfere with the cowl ventilator or operation of the pedal controls. The tubes should be tested and placed in the chassis before it is mounted, as it is difficult to install the tubes after the receiver chassis is permanently mounted. The speaker may be located in any position that is convenient. It is generally mounted on the dash under the cowl. If the set is installed in a tight corner, be sure the antenna trimmer screw is adjusted before the set is mounted. Complete information on this adjustment is given on page 11.

A radio service man and an automobile mechanic or ignition man can co-operate to advantage in installing the receiver. A garage is the best place to work and a portable electric drill, pliers, soldering iron and solder, small wrenches, and a screw driver, are the essential tools.

Although the installation of the automobile radio may appear to be a difficult job, by giving a little thought to the installation before hand as outlined in the foregoing, and with the proper tools the installation is comparatively simple. The performance of an auto radio can hardly be compared with the performance of an AC receiver in the home. The auto receiver must operate on a very small antenna and under many varying conditions. If care is taken, however, in the installation, the performance of this sensitive superheterodyne will be found to be excellent. *Remember, much depends on using the best antenna possible.*

General Procedure

Before installing the receiver look over the units and check them against the parts listed in the back of this manual, so that you are certain you have all of the necessary units.

The general order in installing is to mount the control unit, chassis, flexible drive shaft, speaker, "B" eliminator or "B" battery box and antenna Install the suppressors and condensers for the elimination of ignition and generator noise.

Important

Before installing the control unit, chassis, and flexible drive shaft, read over the data in this manual explaining the mounting of these units, and determine which installation is to be used according to the space available. The important item to be considered in this connection is the distance between the control box and the chassis as there are only three lengths of control shaft which can be supplied, fourteen inches, thirty-four inches, and forty-five inches, and the control shaft cannot be cut

WELLS - GARDNER COMPANY

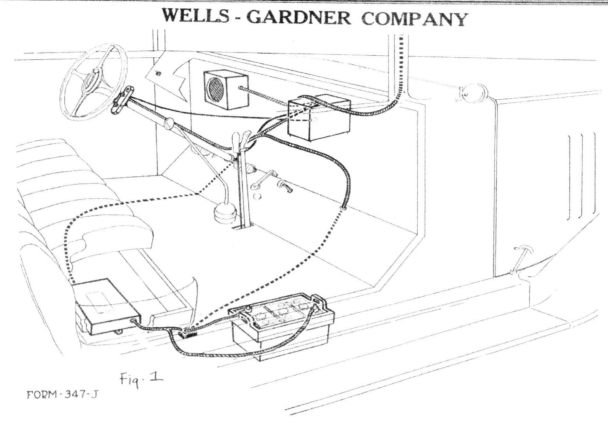

Fig. 1

FORM - 347 - J

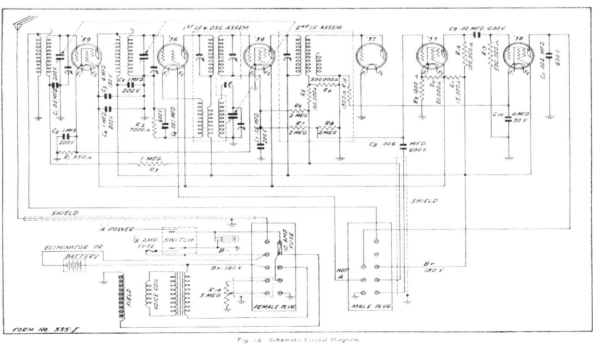

FORM NO. 535-J

Fig. 15 Schematic Circuit Diagram

WELLS - GARDNER COMPANY

Mounting the Control Unit

The control unit is mounted on the steering column under the steering wheel as shown in Fig. 2. The bracket can be screwed to the top and center hole or the center and bottom hole on the left side of the box, depending on the position of the control unit desired.

Wrap one or two pieces of the felt provided around the steering column, leaving room for the set screws to pass before connecting the two parts of the clamp together. The four 8-32 x ⅜" fillister head screws are used for this purpose. When the clamp is in place, take the two 8-32 headless cup

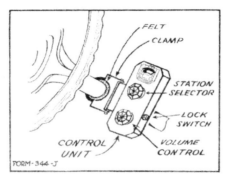

Fig. 2—Control Unit on Steering Column

point set screws and screw them down on the steering column through the holes in the clamp.

The control unit is generally about 4" below the wheel, but this will vary with individual cases. The length of the drive shaft and interference with driver's legs will also govern the location of the control unit.

There are two screws which hold the inside portion of the clamp to the bracket on the box. By loosening these two screws, the box can be swung around if such a position is handier from the standpoint of the person operating the set.

Mounting the Chassis

There are three general ways to mount the chassis as shown in Fig. 3: on the steering column, No. 1, in back of the dash. No. 2, and in front of the dash, No. 3. There are three flexible drive shaft lengths: 14", 34" and 45". The 34" length is regularly supplied with the set unless otherwise specified. The shorter and more direct the flexible drive shaft is, the easier it will turn.

Mounting Chassis on Steering Column

Mounting the chassis on the steering column is by far the easiest method, but be sure there will be no sharp curves in the drive shaft or this method cannot be used.

A steering column mounting is provided and is composed of two parts: the base and the clamp. First attach the base to the chassis box. There are four brackets on the bottom of the chassis box to which the base is attached. It will be noted that the base can be put on lengthwise or crosswise of the bottom of the chassis.

The chassis may be mounted over, or on the side of the column, depending on the space available. It should be mounted in such a way as to make the flexible drive shaft to the control unit as short and in as straight a line as possible. The chassis should not interfere with the feet or legs of the driver, nor with the action of the pedals, hand brake, cowl ventilator or any other apparatus.

Secure the steering column mounting base to the chassis brackets with four of the 10-32 x ⅜" fillister head screws. The other six screws of this type supplied are used to screw the clamp of the steering column mounting to the base. Two or four of the pieces of felt provided should be wrapped around the steering column before the mounting goes on. When the mounting is in place, take the two ¼" No. 20 Cup Point set screws and screw them down on the steering column through the holes in the clamp.

Before the chassis is permanently mounted, the tubes should be inserted, antenna trimmer adjusted (as explained in section on trying out the set), and the flexible drive shaft connected (as explained in next article).

Mounting Chassis in Back of Dash

If the chassis cannot be mounted on the steering column the next best place is in back of the dash, position 2, Fig. 3. Locate it in such a way that the flexible drive shaft to the control unit will have as few bends as possible. In general the 34" length will be used for this method of mounting. Well up under the cowl and to the right of the steering column is a good location.

First drill the three mounting holes required for the dash mounting plate. The location and size of these holes is shown in Fig. 4. A template for drilling these holes is supplied with the set. Three 3" square head mounting bolts are supplied. Take two

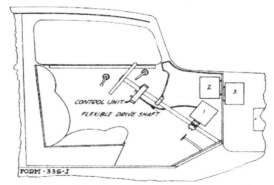

Fig. 3—Possible Chassis Locations—Position 3 used only when absolutely necessary.

WELLS - GARDNER COMPANY

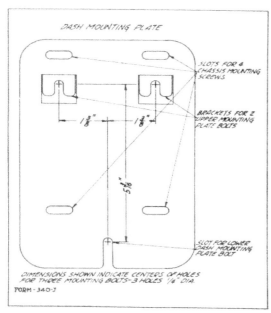

Fig. 4—Dash Mounting Plate

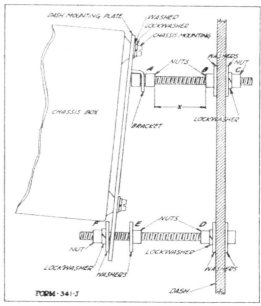

Fig. 5—Details of Chassis Mounting on Dash

of these, which will be used for the upper part of the mounting plate and screw on nut "A" (See Fig. 5). The nut should be just far enough away from the head of the screw to permit the bracket of the mounting plate to slip down as shown in the illustration. Then put on nut "B" and a washer, after which the two bolts can be put through the dash, with the shanks extending into the engine compartment as shown in Fig. 5. A washer, lockwasher, and nut are then put on these bolts, from the front of the dash to hold them in place.

NOTE: If the chassis is mounted with the cover on the bottom, it will be necessary to drill the lower mounting hole $5\frac{1}{8}$" from the top mounting holes rather than $5\frac{3}{16}$" as shown in Fig. 4. Also, it will be necessary to put several washers between the dash mounting plate and the lower mounting holes on the chassis box, indicated in Fig. 4. The latter is necessary in order to keep the dash mounting plate from interfering with the wing nuts if the cover of the chassis box is taken off.

The distance "X" between nuts "A" and "B," which determines how far out the chassis is mounted from the dash, will vary with the model of car. If there is a lot of apparatus in back of the dash, such as wires, tubing, etc., the chassis will have to set out far enough to clear it. If there is little or no intervening apparatus, the chassis can be set in closer to the dash. In general, get it as close as possible. Then put a washer on the third mounting bolt and put this bolt through the lower hole with the head on the engine side of the dash as shown in the illustration. Put on a washer, lockwasher, and nut "D" and tighten it up. Then put on nut "E," screwing it on

far enough so that it will not interfere with the mounting plate.

Next, secure the dash mounting plate to the chassis box by means of the four chassis mounting screws. Note that there are four screws on one of the narrow sides of the box and four screws on one of the broad sides. The purpose of this is to permit the attachment of the plate to whichever side is most convenient. Consideration should be given to the space available and also to the location of the anchor bushing on the chassis box. In general, the cover of the chassis should be at the bottom in order to get at the tubes. All the tubes should be in the sockets and the antenna trimmer adjusted (as explained later) and flexible drive shaft connected before the chassis is permanently installed.

The four mounting screws pass through the four slots in the mounting plate. After they are in place and tight, the dash mounting plate with chassis attached is slipped over the three mounting bolts. The two upper brackets on the plate slip down in back of nut "A" as shown in Fig. 5, and the slot at the bottom of the plate slips over the shank of the lower bolt in back of nut "E." The plate will then hang with the bottom farther away from the dash than the top. A washer, lockwasher and nut "F" are then put on the lower mounting bolt. Nut "F" is then screwed on until the mounting plate is about parallel with the dash. In this position, the bracket at the top of the mounting plate should butt up against nut "A" and be tight. If it is not, continue to screw on nut "F" a slight amount. Nut "E" can then be screwed back and tightened against the mounting plate.

WELLS - GARDNER COMPANY

Mounting Chassis in Front of Dash

This position of mounting should be used only if the other two locations are not possible. Mounting the chassis in front of the dash is undesirable because interference from the car ignition system is greater, the set may be ruined by water and the cable must be unsoldered to get it through the dash.

In general, the procedure is the same as described for mounting in back of the dash. The chassis should be mounted with the anchor bushing on the side so that only a 90° bend is necessary to bring the flexible drive shaft through the dash. When mounted in front of the dash the chassis cover should be on top to get at the tubes.

Attaching the Drive Cable

As already mentioned, the flexible drive shaft comes in three lengths: 14″, 34″ and 45″. The 34″ length is supplied unless otherwise specified on the order. The other lengths may be had by special order or by so specifying at the time the order for the set is placed. *The shaft cannot be cut to length.*

If the 14″ length cannot be used, the next best length, of course, is the 34″ length. The chassis must be so placed relative to the control unit that this length of flexible drive shaft can be put on with a minimum amount of bending. In general, one large radius 90° bend or an easy spiral around the steering column is all that is necessary. The less the number of bends and the larger the radius of them the easier the drive will turn.

Attach the flexible drive shaft at the control unit first. Take off the bottom portion of the box by removing the station selector knob and unscrewing the end screws. The bottom portion of the box may then be dropped away as far as the leads will permit.

In Fig. 6 are shown the constructional details of the flexible drive shaft connections. First loosen the clamping nut on the anchor bushing. Pull the end of the drive shaft about 1½″ out of the casing and push it into the hole at the center of the drive pinion. There is a set screw in the pinion which holds the drive shaft in place. When the shaft is inserted the flat portion should be under this set screw. Tighten down the set screw on this flat portion.

Then push the flexible drive shaft casing into the hole in the anchor bushing and tighten down the clamping nut. This presses the slotted sections of this bushing down on the casing, holding it firmly in place. *Do not tighten the clamping nut excessively.*

Check the centering of the anchor bushing with relation to the holes for flexible shaft. If the end of the casing presses against the shaft it will turn hard. Check all moving parts for grease and apply some if necessary.

The same procedure is then followed in attaching the flexible drive shaft and casing at the chassis. The dial scale should be at the low frequency end stop when the rotor plates are completely meshed. Calibration is very simple on this model and is very easily accomplished after the drive shaft is installed, by continuing to turn the station selector knob at one end of the scale or the other until the scale is correctly set.

If the stops on the dial gear in the control unit act before the stops on the drive gear on the condenser rotor, it will be necessary to loosen the set screw on the bushing of the drive gear rotor. Shift this gear in a counter-clockwise direction the amount necessary to bring the gear stop into action at the same time as the control unit gear at the high frequency end. When this has been done, the gang condenser will act as its own stop at the low frequency end and the gear as the stop at the high frequency end.

The complete assembly should be tried out before the chassis is permanently fastened.

Before tightening the clamping nut on the casing at the chassis, loosen the clamping nut at the control box end. Then adjust the casing until it is securely clamped at both ends when the clamping nuts are tightened down. The flexible drive shaft may, if desired, be taped to the steering column and clamped to the dash.

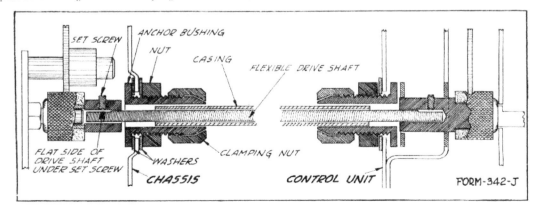

Fig. 6—Details of Flexible Drive Shaft Connections

WELLS - GARDNER COMPANY

Mounting the Speaker

An electrodynamic speaker installed in a wood case is supplied. Acoustically, the best position for the speaker has been found to be on the dash as shown in Fig. 1. Mount it as low as convenient. It may be mounted over the steering column as in the illustration or at any other convenient position on the dash. Before mounting it on the extreme right side of the dash, consideration should be given to the possibility of a car heater being installed. It is not advisable to mount speaker very close to the chassis as in some cases microphonic noises will result.

Before proceeding with the mounting of the speaker, connect the speaker cable to the terminal strip. The shielded four-lead cable passes through the hole on one side of the box. Connect the cable to the terminal strip on the speaker as explained in the section on wiring.

The tone control is mounted on the speaker. Mount the speaker in such a position that the knob will be most accessible.

Mounting "B" Eliminator and Relay

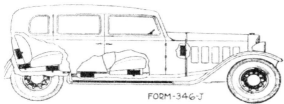

Fig. 7—"B" Eliminator Locations

In addition to the following instructions, a complete installing bulletin for the "B" eliminator is furnished by the manufacturer with each unit. The "B" eliminator can be conveniently mounted in a number of locations in the car as shown in Fig. 7. Under the front seat or in the motor compartment under the hood is a convenient place. The eliminator should be at least 12" away from any ignition or lighting wires of the automobile. Never install the eliminator on end, that is, with the mounting brackets at the top and bottom. Short out the "B" fuse when a "B" Eliminator is used.

In Fig. 1 the "B" eliminator is shown under the front seat, at the right hand side, for illustrative purposes. If, as shown in the illustration, the antenna lead comes down the right front corner post and the "B" eliminator is under the front seat, it should be moved to the left as far as possible. In general, mount it on the opposite side of the car that the antenna lead is installed.

The relay should be mounted near the car storage battery so that the two leads will reach. It is mounted on the frame of the car. Before making any connections to the battery, determine which side is grounded and which side is ungrounded. Then find out if the ungrounded or hot side is positive or negative. This will vary with the make of car.

In Fig. 8 is shown how the connections are made in either case. Unscrew the clamp bolts on the battery and connect lug of yellow lead to the "hot" side of the battery and the lug of the black lead to the grounded side. The bolt goes through the hole in the lug and the lug is bent over. Connect the shielded two-lead cable from the "A" battery and relay to the "B" eliminator. Note that the proper connections will depend on which side the battery is grounded. The "B" cable connections from the chassis may then be completed to the "B" eliminator. It is important that the "B" cable to the eliminator be located as far away from the "A" supply cable as possible. Run them to the "B" eliminator at opposite sides of the car as shown in Fig. 1.

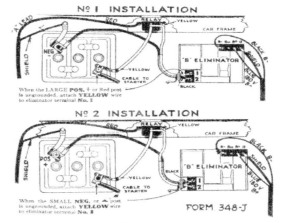

Fig. 8—"B" Eliminator Connections

Testing With "B" Batteries

If for any reason the set should be tested with "B" batteries, the diagram shown in Fig. 11 should be followed. Because of the extremely short life of "B" batteries on automobile sets, they are not recommended for permanent installations. The "B" eliminator is far more satisfactory, less trouble to install and is much cheaper in the long run. The occasion might arise, however, when it is desirable to use "B" batteries for test purposes to determine whether the "B" eliminator is performing properly or not.

WELLS - GARDNER COMPANY

Installing Antenna

First see if there is a built-in antenna. Many cars today come equipped from the factory with a roof antenna. The lead-in generally goes down to the right front corner post and is up under the cowl at the right hand side (facing forward). This lead is connected to the white antenna lead from the set. Care should be taken not to have the lead come in contact with the shield on the antenna lead from the set. Ground the shield on the antenna lead-in at the antenna end.

For any type of antenna, keep the lead-in as far as possible away from the "B" eliminator and from the car ignition system. To try out the effectiveness of any antenna used, check the volume against the volume when using a straight length of wire about 15 feet long run out of the car through one of the windows. If there is no built-in antenna, one of the following can be installed.

Remember, the better the antenna the better will be the reception.

Roof Antenna

The built-in roof antenna is the most satisfactory type. To get inside of the top, it is advisable to employ the services of an experienced man. Otherwise the top may be severely damaged. Most tops have a chicken-wire mesh which is used to support the roof material. It will be necessary to determine if this screen is grounded. To do this, use a continuity meter. By means of a wire, attach a darning needle to one of the prods, poke the darning needle into the roof material, and turn it around until it comes in contact with the chicken wire. Then ground the other prod and if the continuity meter shows a complete circuit, the chicken wire mesh is grounded.

It will be necessary in a case of this kind to remove the top material and cut away the chicken wire from the side supports until it is at least 3″ away from ground at any point. It should also be at least 3″ away from the dome light and the dome light wiring. The chicken wire may then be laced to the points from which it was cut with a heavy, waxed cord.

The chicken wire will then make a satisfactory antenna, or a copper screen may be used. A piece of copper screen at least four square feet in area will be sufficient. Use shielded wire for the lead-in and bring it down the right or left front corner post depending on set location. A piece of loom should first be put over the lead-in and the shield placed over the loom so as to reduce the ground capacity.

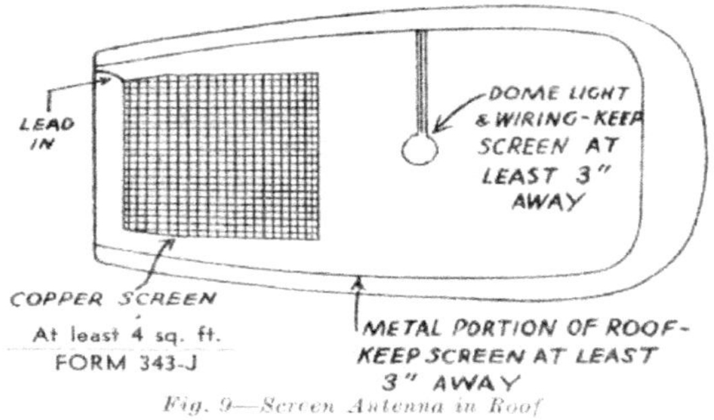

Fig. 9—Screen Antenna in Roof

Tape Antenna on Roof

The tape antenna on top of the car roof is one of the easiest types to install, but it is not recommended as it does not stay on permanently. It is not satisfactory if there is a grounded chicken wire mesh in the car roof as explained in the previous section.

The tape antenna consists of a tinfoil covered tape which comes in rolls. A lug is placed in the outside end to which the lead-in wire is soldered. Unroll the tape at the hole through which the lead-in wire will go and back on the car roof for three or four lengths. Adhesive tape should then be laid over the tape antenna and top dressing or shellac brushed over the adhesive tape. This will help to keep it in place.

Plate Antenna

There are a number of plate antennae on the market at the present time. In general, this type of antenna is not satisfactory and should be used only if no other type of antenna can be installed. The plate antenna generally consists of a metal plate, 2′ to 3′ long, suspended under the running board and attached to the running board by insulators, 2″ to 4″ long. The plate may also be suspended from the channel frame of the car by means of insulating mountings. The lead-in is brought up to the chassis in such a manner as to avoid the car ignition wires as much as possible.

Under-Car Antenna

The under-car antenna is considered more satisfactory than the plate antenna. The under-car antenna consists of a wire fastened from the right side of the rear axle to the lowest point under the motor, then back to the left rear axle, forming a V. At the vertex of the V is a spring to take up the slack. The lead-in is brought up from the vertex end.

Wiring

After all units have been installed, the cable wiring can be completed. In Fig. 10 is shown the complete wiring diagram. "B" batteries are shown. The proper connections for a "B" eliminator are shown in Fig. 8. CAUTION—Do not turn set on until all wiring connections are completed.

Note that there are four shielded cables from the cable head of the chassis. One of these goes to the control box. Put the cable head in place on the chassis temporarily and fasten this cable to the dash and steering column. Extending from this cable is the shielded "A" lead which goes to the "hot" "A"

WELLS - GARDNER COMPANY

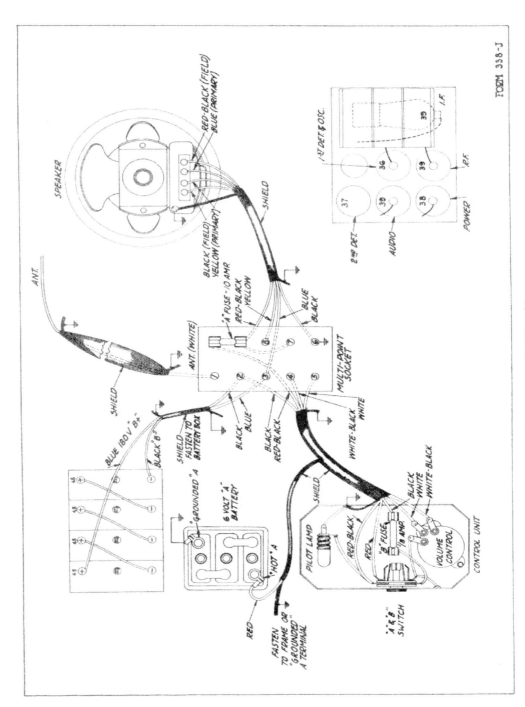

FORM 338-J

Fig. 11—Complete Unit Wiring Diagram

WELLS - GARDNER COMPANY

connection on the battery or to the relay, when a "B" eliminator is used. When making this connection, be sure that the grounded shield does not short to the "hot" "A" terminal. The "A" lead should be as short as possible.

The speaker cable, the antenna lead, and the "B" cable are connected to their respective units. If any of these leads are too long they may be cut to length. All shields should be well grounded at both ends, generally to the case or the frame of the unit to which they are connected.

The shield on the speaker cable is grounded to the screw adjacent to the speaker terminal strip as shown in Fig. 10. If the shield pig-tail is too long, cut it short before making the connections in order to keep it from shorting out any of the speaker terminals. In the case of the "B" eliminator installa-tion, keep the "B" cable as far as possible away from the "A" cable and all "A" connections. If the "B" battery lead must go under the car, cover it with the piece of loom supplied, to keep out moisture and to prevent the shield from rattling against the car body. In a "B" eliminator installation, ground the "B" cable shield to the "B" eliminator box.

The antenna cable should run up behind the in-strument panel and directly over the point where the aerial lead-in comes in. The lead-in wire should be as short as possible. When connecting the aerial lead from the set to the lead-in wire from the an-tenna, be sure that neither of these two wires touches the grounded shield.

The shield of the antenna lead must be well grounded at the antenna end to the nearest con-venient point on the chassis or metal portion of body.

Trying Out the Set and Adjusting

After the wiring has all been completed and be-fore the chassis is permanently installed, insert the tubes, try out the set, and adjust the antenna trim-mer condenser. The tube location is shown in Fig. 10. Put one of the rubber bands around each tube. Do not start the engine of the car.

To adjust the antenna trimmer, tune in a weak signal at the high frequency end of the dial with the manual volume control $3/4$ on. On one end of the chassis box is a small metal plate. Remove the two screws holding this plate. Directly under the hole in the chassis box is the antenna trimmer condenser screw. Turn this adjusting screw up or down until maximum output is obtained.

If the receiver does not work, check the "A" and "B" voltages. CAUTION—These voltages should be checked only at the sockets in the receiver or at the "A" and "B" units. Do not check the voltages by removing the cable head and reading them at the multi-point socket. The reason for this is that if the switch is turned on and off with the multi-point socket not connected, the pilot light lamp may be burned out, due to the inductive surge caused by the speaker field. ALWAYS have the multi-point socket in the cable head inserted and all connections com-pleted before turning the switch on or off.

Suppression of Ignition and Generator Noise

After the receiver is in satisfactory working order, start the motor and note the amount of noise. As a general rule, spark plug suppressors, a distributor suppressor and a $1/2$ mfd. condenser on the genera-tor are all that is required for the reduction of ig-nition and generator noise. If these items do not reduce the noise sufficiently, other measures as de-scribed below are required.

One spark plug suppressor is required for each plug. The method of mounting is shown in Fig. 12. Remove the wire from the top of the plug, put the suppressor on, and attach the wire to the top of the suppressor.

A distributor suppressor is put in the high ten-sion lead, between the coil and the distributor head. Position "C," Fig. 12, on the distributor head is the most satisfactory and most commonly used point of mounting. If this is not practical, the high tension line may be cut close to the distributor head and the distributor suppressor with wood screw ends in-serted in the line as shown in position "B."

The $1/2$ mfd. generator condenser is installed as shown in Fig. 12. The lead from the condenser goes to one side of the cut-out connection on the genera-tor. The mounting clamp grounds the other side of the condenser.

After the above procedure has been followed, again start the motor. If noisy operation persists, a number of steps can be taken and the various sug-gestions as given can be tried until the noise is sat-isfactorily reduced.

Try two suppressors in the high tension line, one at the coil end in addition to one at the distributor end, position "C," Fig. 12.

Ground all cables and tubing which pass through the dash, such as oil lines, gas lines, etc. Ground to the dash or at the nearest convenient point on the frame with a good short ground connection. Use the left-over shield from the "B" battery lead for this purpose.

If the chassis and coil are both in back of the dash (under the cowl), take off the coil and mount it on the front of the dash (in the engine compartment). If the coil cannot be moved, place a copper can over it and ground the can at the coil mounting.

Clean and respace spark plugs—clean and check distributor points—check distributor condenser.

In some cases, the high and low tension leads be-tween the coil and distributor are run close to-gether. In some cases they are in the same conduit. If this is the case, remove the low tension lead from this conduit. In any event, keep the high and low

WELLS - GARDNER COMPANY

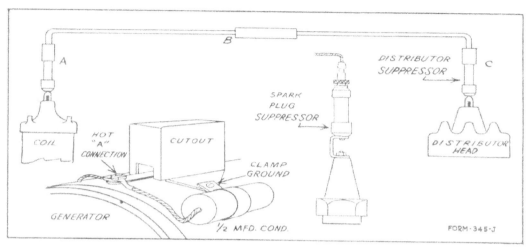

Fig. 12—Installation of Suppressors and Condenser

tension leads as far apart from each other as possible. Shield and ground the high tension lead, if separating the two leads is not sufficient. Then try also shielding the low tension lead.

A .5 mfd. condenser is necessary in some cases between the low tension lead terminal on the coil and ground. In other cases, this condenser might be harmful. It can be tried out, however, experimentally.

In some instances it will be helpful to connect a generator condenser from the dome light wire at the terminal block on the dash to the ground.

Noise, on occasion, may be due to weak pickup caused by a poor antenna. The action of the automatic volume control, due to the low pickup, causes the set to operate at maximum sensitivity, thereby increasing noisy reception, due both to external pickup and internal conditions.

Noisy operation is also caused in some instances by loose parts in the car body or frame. These loose parts rubbing together affect the grounding and cause noises, due to the rubbing or wiping action. Tightening up the frame and body at all points, and in some cases, the use of a copper jumper will eliminate noise of this nature.

Noise may also be due to the "B" eliminator. Keep the eliminator as far away from the receiver and lead-in as possible. Also, ground the case. The eliminator can cause hum if the filters are defective or high frequency energy can be radiated directly from the case.

Be sure there are no loose lights or wiring.

Care and Maintenance

Advancing Generator Charging Rate

The installation of the automobile radio imposes an additional drain on the car storage battery. This can be compensated for by advancing the charging rate of the car generator. Check the state of charge of the storage battery about a week after the installation of the automobile radio is made and adjust the charging rate accordingly.

Tubes

The type of tubes used and location of these tubes in the chassis are shown in Fig. 10. These tubes are designed especially for auto receivers. Most of them, under normal usage, will last for many months and in some cases, years. Some of them, however, may become faulty after a few months of operation.

For that reason, try out a new set of tubes periodically, inserting them in the receiver one at a time and noting any difference in performance.

Fuses

Two fuses are used on this receiver. One for the "A" line and one for the "B" line. As shown in Fig. 10, the "A" fuse is a 10 amp. fuse and is located on the multi-point socket. The "B" fuse is a $\frac{1}{8}$ amp. fuse and is inside of the control unit.

To change the "B" fuse it will be necessary to remove the cover of the control box, and to change the "A" fuse it will be necessary to take the cable head off the chassis. Be sure that the switch is off when changing fuses.

Pilot Lamp

The pilot lamp is a standard six-volt No. 40 lamp. To replace the lamp, remove the cover of the control unit. The bottom portion of the box will now drop away as far as the leads will permit. The light socket clip and lamp can then be easily removed by first removing the dial which is held by one screw in the center.

WELLS - GARDNER COMPANY

"B" Eliminator or "B" Batteries

The voltage of the "B" eliminator should be checked occasionally with a high resistance voltmeter. The tube in the "B" eliminator may burn out after three to nine months' use. If the eliminator is of the rotating type, the bearings will require oiling periodically.

If four 45 volt "B" batteries are used for the "B" supply, these will run down after two to five months, depending on the amount the set is operated. When the voltage of a battery drops below 30 under load, a new one should be purchased.

Electrical Condition of Car

Dirty spark plugs, incorrect spacing of distributor points, faulty distributor condenser, and various other items in the car electrical system can cause noisy operation. If the customer complains of noise in the receiver after it has been in use for some time, check the items mentioned as well as other parts of the car electrical system for poor connections, grounds, and other faults which may be responsible for the noise.

Keep Units Dry

Caution the customer, when having the car washed, to avoid getting the chassis and "B" battery box or "B" eliminator water-soaked. Water getting into these units may cause damage and deterioration and, in some cases, a short circuit. Driving the car through an excessive amount of mud or water may bring about the same result.

Circuit

The circuit consists of an antenna stage, a '39 R. F. stage, a '36 Detector-oscillator stage, a '39 I. F. stage, a '37 diode detector stage, a '39 first audio stage, and a '38 output stage.

The intermediate frequency is 262 K. C. The diode current establishes a drop across a resistor network, which is used as an additional bias voltage on the R. F. '39, I. F. '39, and audio '39 tubes, giving automatic volume control action.

The full control voltage is supplied to the R. F. tube, two-thirds to the I. F. tube, and one-third to the audio tube. As the signal increases in intensity, the applied control voltage is increased, thus giving uniform output as set by the manual volume control. The manual volume control varies the diode audio voltage applied to the first audio tube.

An electrodynamic speaker with the field energized by the six-volt car battery is used. Power for the receiver is obtained from the car storage battery and from a "B" eliminator or from "B" batteries. The tone control is mounted on the speaker. The tubes used are the new six-volt tubes especially designed for automobile radio receivers.

Voltages at Sockets

In the following chart are given the voltages at the sockets. Before checking the voltages at the sockets, a convenient point, in some cases, to check the applied "A" and "B" voltages is at the speaker terminal strip. A high resistance voltmeter should be used.

CAUTION—Do not check the "A" and "B" voltages at the multi-point socket on the cable head, as the pilot light may be burned out when the switch is turned off. This is due to the high inductance of the speaker field, which will increase the voltage at the break of the circuit. Also, when the cable head and multi-point socket is taken off, the connections between the chassis and power unit are open so that readings are not made under load conditions.

To read the voltages at the sockets, the chassis box, in most cases, will have to be taken off of its mounting. In some instances, the cables, which may be attached to the dash or at other points, will have to be taken off. The voltages can be read at the sockets with a long plug or with a pair of long, insulated test prods. If these are not available, it will be necessary to remove the chassis from the box. The multi-point socket on the cable head is then re-connected to the multi-point plug on the chassis. Considerable care must be taken when the chassis is out of the case in this manner to prevent accidental short circuits of plus "B" or plus "A" points to ground.

All tubes must be inserted and all units connected. A signal will effect the control voltages on the R. F., I. F., and first audio tubes. If signals are received, ground the antenna and remove the second detector tube to make the other readings.

Type of Tube	Function	Across Heater	Plate to Cathode	Screen to Cathode	Grid to Cathode	Normal Plate M.A
'39	R. F.	6.	177	80	3	3.6
'36	1st Det.	6.	173	76	7[1]	.9[1]
'39	I. F.	6.	177	80	3	3.6
'37	2nd Det.	6.	0		0	0
'39	1st Audio	6.	54	77	6	1.2
'38	Output	6.	159	165	15.5	10

[1] Will vary with dial setting.

NOTE: All bias voltages must be read from cathode to ground.

WELLS - GARDNER COMPANY

No. Z6Z1 Series
UNIVERSAL MODEL
Automobile Radio
INSTALLATION AND SERVICE MANUAL

Description

The Series Z6Z1 receiver is a completely self-contained auto radio receiver designed for quick and simple installation at several locations in the automobile. It may also, if desired, be used with a separate steering column control unit. When mounted this way, two flexible shafts mechanically connect the control unit to the chassis. One of these is for the volume control and switch and the other is for the tuning mechanism. A roof antenna is recommended. Current to operate the receiver is obtained from the car storage battery.

In this manual are covered detailed installation instructions and information for completing and maintaining the installation. Data for servicing the receiver is also included, should the necessity for such procedure arise. The following tools are required: portable electric drill, screw drivers, pliers, a heavy soldering iron, hack saw, files, small wrenches, and cutters.

Before making the installation it is suggested that this manual be completely read.

Integral Mounting of Chassis

By integral or all-in-one mounting of the chassis is meant operating the receiver by means of the controls on the chassis box (and not with a separate control unit). This method is the simplest, as no changes are required on the receiver. It can be installed in several ways, as explained below and as illustrated in Fig. 1. Still other methods of mounting and locations for the chassis will suggest themselves, depending on the space available and variations in the construction of different cars.

Floor or Shelf Mounting

In Fig. 1(A) is shown how the chassis can be placed on the floor in front of the front seat. There are four rubber mounting feet on the bottom of the box, on which it stands. It may also be placed in back of the front seat (B) so as to be in the rear compartment of the car. In some cars, there is room enough between the two front seats for the chassis box to be placed. In coupes, the chassis may be placed on the shelf in back of the seat. Still other locations, as mentioned above, can be used, depending on the space available in different cars.

After the position is decided on, the chassis is permanently mounted in place by means of the two case mounting feet supplied for this method of mounting. These mounting feet are shown in Fig. 1. One side of the foot, which is a small angle bracket, is secured to the end of the chassis box by means of one of the chassis mounting screws. The other side of the foot is screwed to the floor board or surface on which the chassis is resting, with a wood screw. The two feet are placed diagonally, that is on one end of the chassis box it is at the front, while on the other end it is at the rear.

Flush Mounting of Chassis

In Fig. 1(C) is also shown how the chassis can be mounted on the dash by means of brackets, in such a way that the front portion of the box with the controls, is flush, or nearly so, with the instrument panel. This is a very desirable method of installation, as the receiver is rigidly in place, out of the way, and the controls are very accessible.

When mounted this way, two side case brackets (long type) are used, one on each end of the box, as shown in Fig. 1. Two mounting screws are generally used to secure each bracket to the end of the chassis box. Three may be used in cases where the distance between the instrument panel and dash is small. Six embossings with inset nuts are provided on each end of the chassis box. Any two of these or

WELLS - GARDNER COMPANY

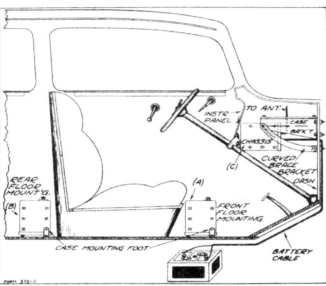

Fig. 1—Integral Mounting—Side View

interference with the legs of the driver or passenger in the front seat and also to the possibility of interference with the controls of the car, such as pedals, gear shift lever, and hand brake lever, before the location is definitely decided on. The possibility of a car heater installation may also be considered. After the location is decided on, drill the four mounting holes required. The location and size of these holes is shown in Fig. 3. A template for drilling these holes is supplied with the receiver. Six 1/4" mounting bolts, six washers, six lockwashers and six nuts are provided. The mounting bolt is put through the bracket and dash with the shank extending into the engine compartment. A washer, the lockwasher and nut, are then put on. Mount the brackets permanently, but do not mount the chassis permanently until the wiring connections are completed, the tubes are all inserted, the receiver tried out, and the antenna trimmer adjusted (explained later).

When the case brackets are in place, the curved brace brackets can be installed. These can be put on in a number of different ways. The front or back case bracket screw can be used and the brace bracket itself can be mounted upward or downward. As a general rule it is mounted on the bracket screw farthest away from the dash and downward as shown in Fig. 1. The small angle brackets supplied with the receiver are secured at the base of the curved brace brackets as shown in Figs. 1 and 2, by means of the No. 10-32 3/8" Round Head Screw, nut and washer supplied. After the position of the brace brackets is decided on, put them in place and start the holes for them with a center punch. These brackets are bolted to the dash in the same manner as explained above for the case brackets.

three, as mentioned above, may be used for the bracket screws, which, together with the slots in the brackets, provides great flexibility in mounting. In addition to the side case brackets, two curved brace brackets and one cross strap brace as shown in Figs. 1 and 2 are used.

The chassis should be mounted as close to the center of the instrument panel as possible. This makes the controls accessible to people in either front seat. As stated above, it should be mounted so that the front side of the box with the controls, is flush or nearly so with the instrument panel of the automobile. If car apparatus or space available prevent the mounting of the chassis at the center,

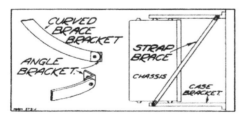

Fig. 2—Angle Brackets and Strap Brace

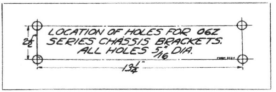

Fig. 3—Mounting Hole Location

it may have to be moved to either side. In some instances, it can be mounted at the center of the instrument panel, but may have to be moved down and nearer to the dash than as shown in Fig. 1. Consideration should be given to the possibility of

Next, put the strap brace in place. This is mounted diagonally across the two brace brackets as shown in Fig. 2. There is a tapped hole at either end of the top flange of the case brackets which are used for this purpose. Two 10-32 1/4" long bolts are provided for the strap brace.

WELLS - GARDNER COMPANY

Separate Control Unit Mounting of Chassis

In this method of mounting, the chassis is mounted on the dash and is operated from a separate remote control unit which is on the steering column. Two flexible shafts mechanically connect

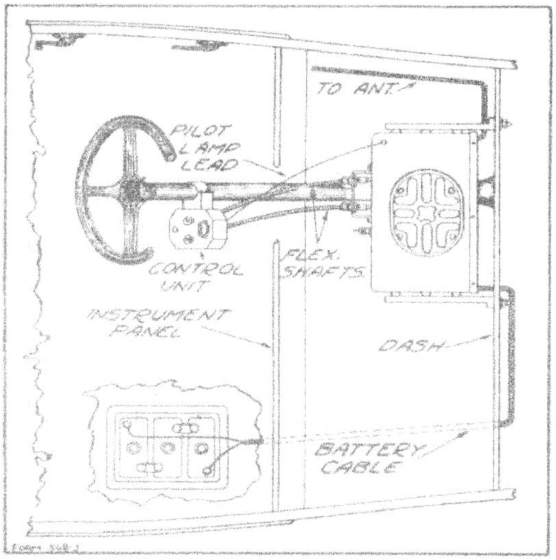

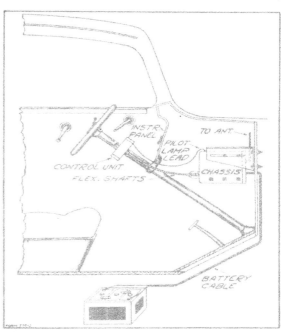

Fig. 4—Chassis with Control Unit—Top View

the control unit and the chassis. This method of mounting is very desirable as the controls are most accessible to the driver. The items required for this method of mounting are shown in the installation list at the back of the manual. The procedure for this method of installation is as follows:

Mounting the Control Unit

The control unit is mounted on the steering column under the steering wheel as shown in Figs. 4 and 5. A clamp is used to hold it in position.

The outer portion of the clamp is screwed to the inner portion by means of the four 8-32x⅜" fillister head screws supplied with the receiver.

Two rubber strips are provided, one ⅛" thick and the other ⅟₆" thick. These are wrapped around the steering column under the clamp. Either or both of these strips may be used, depending on the thickness of the column. Wrap the rubber strips around the column in such a way as to allow the set screws which hold the clamp in position to pass through. When the clamp is in place, take the two 8-32 headless cup point set screws and screw them down on the steering column through the tapped holes in the clamp.

The control unit is generally about 4" below the wheel, but this will vary with individual cases. The length of the drive shaft and interference with

driver's legs will also govern the location of the control unit.

There are two screws which hold the inside portion of the clamp to the control unit swivel. By

Fig. 5—Chassis with Control Unit—Side View

loosening these two screws, the box can be swung around if such a position is handier from the standpoint of the person operating the set. Instructions for attaching the pilot lamp lead are contained in the article "Completing the Wiring Connections."

Mounting the Chassis

The chassis is mounted on the dash by means of two short brackets, as shown in Figs. 4 and 5. Two or three mounting screws are used to secure each bracket to the end of the chassis box. Three are used if the chassis is close to the dash and two if it is set out some distance. In general, keep the chassis as close to the dash as possible. The procedure for attaching the brackets to the chassis box and to the dash is the same as explained above for mounting the side case brackets under the article, "Flush Mounting of Chassis." No curved brace brackets or strap braces are used in this method of mounting.

The chassis should be mounted with the speaker grill facing down and the side with lock and controls facing the listener, as shown in Fig. 4. Before mounting the chassis, the flexible drive shaft con-

WELLS - GARDNER COMPANY

nections as explained in the next article must be made.

The location of the chassis will very often depend on the space available. To the left of the center, as shown in Fig. 4, is a good location. The chassis should be mounted in such a way that the flexible drive shafts to the control unit will be in as straight a line as possible or with large radius bends. *In general, it will be advisable to consider the possibility of a car heater installation at the right side of the dash (facing forward).* In practically every case no difficulty will be experienced in mounting the heater and chassis on the dash. The chassis should be mounted in such a way that the lock *which remains on the chassis box will be accessible.*

The possibility of interference with people in the front seats and with car controls, as mentioned previously, should also be considered.

When the location is decided on, drill the four mounting holes required as shown in Fig. 3 and proceed as explained above. Mount the brackets permanently, but do not mount the chassis permanently until the wiring connections are completed, all tubes are in the sockets, the flexible drive shafts connected, and the antenna trimmer adjusted (explained later).

Attaching the Flexible Drive Shafts

After the control unit is mounted and the chassis is temporarily mounted, the flexible drive shafts may be attached. Two 34" shafts are supplied unless otherwise specified. These shafts may also be had in 14", 20" and 45" lengths.

The flexible drive shafts should always be in-

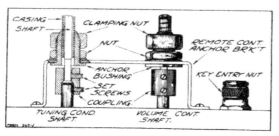

Fig. 6—Details of Flexible Drive Shaft Connections

stalled with a minimum amount of bending. Always keep the radius of the bend as large as possible. The larger the radius of the bend, the easier the shaft will turn.

The 34" shafts supplied with the receiver may be cut to a shorter length if necessary. The shaft (inside portion) should first be brazed at the point to be cut. It should then be cut with a three-corner

file or edge of a grinding wheel. *Do not use a hack saw.* After the shaft is cut, file it down in one place a slight amount to provide a flat surface for the set screw. The casing which is 1½" shorter must be cut to correspond. This should be tinned or brazed first at the point to be cut and may then be cut with a hack saw.

It is advisable to attach the flexible shafts with the chassis on the mounting brackets, but if the chassis is inaccessible, it may be removed from the brackets. Keep it as close to its regular position as possible so that the flexible shaft will not turn after the chassis is replaced on the brackets. In general, it may be moved up or down, but should not be moved sideways or be turned.

To attach the flexible shafts to the chassis, first turn the on-off switch knob to the off position and the station selector knob to the low frequency end stop. Then remove the two knobs. These two knobs are then put on the control unit. Loosen the set screws on the two couplings and slip them over the two shafts as shown in Fig. 6. Then secure the remote control anchor bracket in place on the chassis box by means of the four 6-32-¼" screws. The dial gear and pilot lamp remain in the chassis box.

Next, center the two anchor bushings on the anchor bracket. To do this, first loosen the nut which holds the bushing in place. Center the bushing so that the center of it is in line with the center of the shaft below. Then tighten the nut. Turn the on-off switch and volume control knob on the control unit to the extreme counter-clockwise position. Then extend the volume control flexible shaft into the coupling and tighten the two set screws in this coupling. The outside set screw should be tightened down *on one of the four flat faces* of the shaft. Then tighten down the clamping nut on the volume control shaft casing, *but do not tighten this nut excessively.*

To attach the tuning condenser flexible shaft, proceed in the same manner as above, except that the dial gear in the control unit should first be turned to the low frequency end stop. After the two shafts are connected, mount the chassis in place temporarily if it has been taken off and check the operation of both tuning condenser and volume control. The switch should be off when the volume control knob is in the locked position. It may be necessary to loosen the inner set screw and do a slight amount of adjusting until the proper setting is obtained. In case the dial gear in the control unit is not correctly calibrated or does not coincide with the dial gear calibration in the chassis box, further adjustment of this control can be brought about in the same manner, that is, by first loosening the inner set screw of the coupling. The clamping nut of the tuning condenser shaft anchor bushing is tightened down as explained above.

Antenna

A roof antenna is recommended, as by far the best results will be obtained. A large percentage of

cars at the present time come equipped by the factory with built-in roof antennas. In those cars which

WELLS - GARDNER COMPANY

do not have an antenna, one will have to be put in.

First determine if the top has a grounded chicken wire mesh. To do this, use a continuity meter. By means of a wire, attach a darning needle to one of the prods. Poke the darning needle into the roof material and turn it around until it comes in contact with the chicken wire. Then ground the other prod and if the continuity meter shows a complete circuit, the chicken wire mesh is grounded. In a case of this kind, it will be necessary to get inside of the roof and it is advisable to employ the services of an auto "top man" or an upholsterer.

It will be necessary to remove the top material and cut away the chicken wire from the side supports until it is at least 3" away from ground at any point. It should also be at least 3" away from the dome light and the dome light wiring. The chicken wire may then be laced to the points from which it was cut with a heavy, waxed cord. The chicken wire will then make a satisfactory antenna, or a copper screen may be used.

If the chicken wire is not grounded, it may be used as the antenna by taking down the roof material at one corner and soldering the lead-in wire to it. If it is not desired to take down the roof material a piece of copper screening can be tacked to the roof on the inside of the car. At least six square feet should be used. Keep it at least 3" away from any grounded metal parts on all sides. After the screen is in place, it can be covered over with cloth which matches the roof material. Solder the lead-in wire to the screen and bring it down the front corner post nearest to the set.

Another, and a very simple way in which an antenna can be secured to the inside of the car roof is to use one of the car-roof antennas which are now being made up especially for this purpose. There is one type of antenna which consists of copper strips laid back and forth between two pieces of cardboard. The cardboard is then covered over with material which matches the roof material. This antenna can be had in several colors and is tacked in place on the inside of the car roof in a few minutes.

Completing the Wiring Connections

Antenna Cable

Bring the antenna cable of the receiver in the most direct manner possible to the lead-in from the antenna and connect it to the latter. Keep it as high as possible and as far away from any car wiring as possible. Care should be taken not to have the antenna wire come in contact with the shield wires. Ground the pigtail of the antenna cable shield at the antenna end. The pigtail of this shield at the chassis end is grounded under one of the chassis mounting screws.

In some cases the shielded antenna lead from the receiver is not long enough to reach to the column at which the antenna lead-in comes down. In a case of this kind, cover the exposed portion of the lead-in wire with loom and braided shield from the point where it leaves the column to the point of connection to the antenna lead of the receiver. Connect the two wires together and connect the two shields together, care being taken that no strand of the shield touches the antenna wire.

Battery Cable

The battery cable should be brought over to the storage battery in the most convenient manner possible. In Figs. 4 and 5 it is shown passing through a hole in the dash, thence down and under the floor board to the battery. In other installations, it may be more convenient to bring this cable down in back of one of the side pads and thence to the battery. The lug on the yellow lead of this cable is connected to the "Hot" or ungrounded side of the battery (the "Hot" or ungrounded side may be positive or nega-

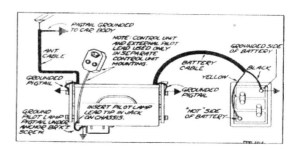

Fig. 7—External Wiring Connections

tive, depending on the make of car). The lug on the black lead is connected to the grounded side of the battery. The pigtail of the shield of this cable at the chassis end should be grounded under one of the chassis mounting screws.

Pilot Lamp (For Separate Control Unit Only)

When a separate control unit is used connect the pilot lamp as follows:

The pilot lamp lead is in a shielded cable which extends out from the control unit box. On the rear wall of the chassis, near one of the ends, will be seen a tip jack. Insert the tip on the end of the pilot lamp lead into this jack. There is also a pigtail or shield extension at the end of this lead. Ground this pigtail with one of the anchor bracket screws (see Fig. 7). Double up the pilot lamp lead if it is too long—*Do not cut this lead.*

WELLS - GARDNER COMPANY

Trying Out the Set and Adjusting

After the wiring has all been completed and before the chassis is permanently installed, try out the set and adjust the antenna trimmer. The location of the tubes is shown in Fig. 8. To adjust the antenna trimmer, tune in a weak signal between 1200 and 1400 K.C. with the volume control about three-fourths on. On one end of the chassis box are two small metal plates. Remove the smaller of these two plates. Directly under the hole in the chassis box is the antenna trimmer condenser screw. Turn this adjusting screw up or down until maximum output is obtained.

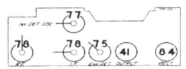

Fig. 8—Location of Tubes

Suppression of Ignition and Generator Noise

Required Procedure

The following procedure will be required in all cases. Distributor Suppressor—Remove the high tension lead to the distributor. Insert a distributor suppressor and connect the wire to the other end of the suppressor (see Fig. 9). If this is not practical, cut the high tension lead *close to the distributor* and use a wood screw end type distributor suppressor in this line.

Generator Condenser—The .5 mfd. generator condenser is installed as shown in Fig. 9. The lead from the condenser goes to one side of the cut-out connection on the generator. The mounting clamp grounds the other side of the condenser. If "Startix" is used try the condenser to both sides of the cut-out to see which way reduces the noise the most.

Dome Light Lead—This lead has been found to be a cause of interference in almost every case and the following steps are therefore put in with the required procedure:

To determine the amount of noise due to the dome light lead, disconnect this lead at the ammeter, block, or where it is connected, coil it up, and tuck it as far as possible up in the column it comes down from. Then, with the engine running, ground the end of this wire. If this is found to reduce the noise noticeably, interference is being radiated by the dome light lead. Reconnect the dome light lead and try a .5 mfd. generator condenser from the connecting point of the lead to ground. If this does not cure the noise caused by this lead, disconnect the lead and encase it in braided copper shield from the point where it leaves the column post to the point of connection. Keep the lead as far as possible away from car ignition wires and ground the shield.

If the noise due to the dome light lead still persists, disconnect the dome light lead and remove it from the front corner post, at which point it is generally run down. Run the lead down one of the side posts in back of the door and direct to the storage battery.

The above steps will, in most cases, reduce the ignition noise to a satisfactory level. It should be remembered that when no station signal is being received, the receiver is operating at its maximum sensitivity, owing to the action of the automatic volume control, and any noise signals picked up will be greatly amplified. As soon as a signal is tuned in, the ratio of the desired signal to the noise signals goes up and the noise is automatically reduced or eliminated. Tune in a distant station and if no noise is received with the signal, the motor noise has been satisfactorily reduced.

Additional Procedure Which May Be Required

If motor noise persists, short the antenna to ground. If this has little or no effect on the motor noise, it is recommended that a set of spark plug suppressors be installed. One suppressor is put on each plug as shown in Fig. 9.

If shorting the antenna to ground stops the noise, the interference is being radiated into the antenna. In a case of this kind, it is reasonably certain that this noise can be eliminated without spark plug suppressors and the following steps can be taken until the noise is reduced:

Put a distributor suppressor in the high tension lead at the coil.

Try a .5 mfd. condenser from the "Hot" side of the coil primary to ground. In some cases, this condenser may not help. It can be tried out, however, experimentally.

Try a .5 mfd. condenser from the ammeter to ground, from the fuse to ground, switch to ground, and various other 6 volt connections to ground and see what effect these condensers have on the noise pickup.

Peen the distributor rotor bar to lessen the gap between the rotor bar contact and points. Clean and respace the spark plugs—check the distributor condenser.

Ground all cables and tubing which pass through the dash, such as oil lines, gas lines, etc. Ground to the dash or at the nearest convenient point on the frame with a good short ground connection. Use braided shield for this purpose.

If the chassis and coil are both in back of the dash (under the cowl), take off the coil and mount it on the front of the dash (in the engine compartment). If the coil cannot be moved, place a copper can over it and ground the can at the coil mounting.

WELLS - GARDNER COMPANY

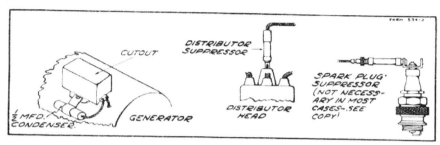

Fig. 9—Installation of Suppressors and Condenser

In some cases, the high and low tension leads between the coil and distributor are run close together. In some cars they are in the same conduit. If this is the case, remove the low tension lead from this conduit. In any event, keep the high and low tension leads as far apart from each other as possible. Shield and ground the shield of the high tension lead, if separating the two leads is not sufficient. Then try also shielding the low tension lead.

The motor must, in every case, be well grounded to the frame of the car. If it is not, use a very heavy braided lead for this purpose, similar to a storage battery ground lead.

Noise, on occasion, may be due to weak pickup caused by the automobile being in a shielded location. The action of the automatic volume control, due to the low pickup, causes the set to operate at its maximum sensitivity, thereby increasing noisy reception, due both to external pickup and internal conditions.

Noisy operation is also caused in some instances by loose parts in the car body or frame. These loose parts rubbing together affect the grounding and cause noises, due to the rubbing or wiping action. Tightening up the frame and body at all points and in some cases, the use of a copper jumper will eliminate noise of this nature.

Care and Maintenance

Advancing Generator Charging Rate

The installation of the automobile radio imposes an additional drain on the car storage battery. This can be compensated for by advancing the charging rate of the car generator. Check the state of charge of the storage battery about a week after the installation of the automobile radio is made and adjust the charging rate accordingly.

Tubes

The type of tubes used and location of these tubes in the chassis are shown in Fig. 8. These tubes are of a sturdy, rugged construction designed especially for an auto receiver. Most of them, under normal usage, will last for many months and in some cases, years. Some of them, however, may become faulty after a few months of operation.

For that reason, it is advisable to secure a new set of tested tubes at intervals of three to six months and have these inserted in the receiver one at a time, noting any difference in performance.

Pilot Lamp

Integral Mounting—A 6-8 volt miniature base pilot lamp is used. When the receiver is operated with the controls on the chassis box, the pilot lamp is inside of the chassis box. To replace the lamp, first turn the receiver off. On the end wall of the chassis box nearest to the dial scale are two metal plates. Remove the larger of these two plates by taking out the screws. In order to get at this plate, it may, when the chassis is so mounted, be necessary to remove the chassis from the mounting brackets. The

pilot lamp socket is secured to a spring clip which is on a bracket on this metal plate. Replace the lamp in the socket and then attach the plate.

Control Unit Mounting—When a separate steering column control unit is used, the pilot lamp is in the control unit. To replace the lamp, first turn the receiver off. Remove the two control knobs and hexagon nut. Then take out the screw holding the control box cover in place, after which the cover can be taken off. The pilot lamp socket is secured to a spring clip which is on a bracket in the control unit. Push this clip and socket over far enough to get at the lamp, after which the bulb can be replaced and the control unit reassembled.

Fuse

A 10 amp. automobile fuse is used for the "A" line. This fuse is mounted on a block on the power transformer in the chassis. To change the fuse, it will be necessary to remove the cover of the chassis box.

Electrical Condition of Car

Dirty spark plugs, incorrect spacing of distributor points, faulty distributor condenser, and various other items in the car electrical system can cause noisy operation. If the customer complains of noise in the receiver after it has been in use for some time, check the items mentioned as well as other parts of the car electrical system for poor connections, grounds, and other faults which may be responsible for the noise.

WELLS - GARDNER COMPANY

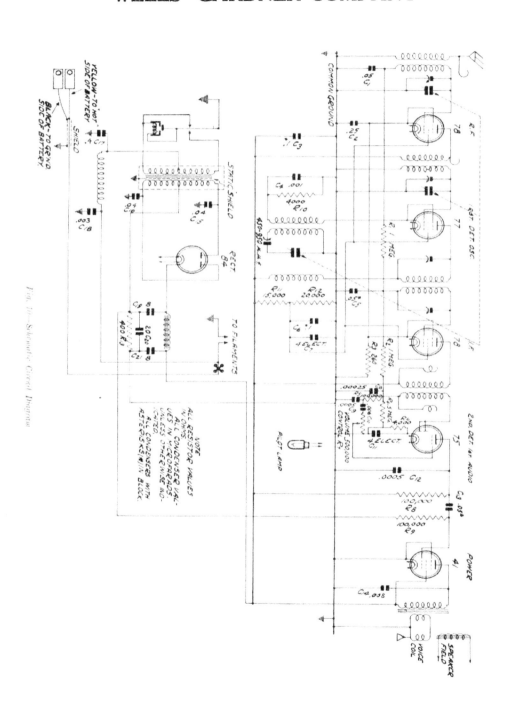

Fig. 19—Schematic Circuit Diagram

WELLS - GARDNER COMPANY

Circuit

The circuit consists of an antenna stage, a 78 R.F. stage, a 77 1st detector-oscillator stage, a 78 I.F. stage, a 75 dual diode-triode tube, which functions as a diode 2nd-detector and triode 1st audio stage, and a single 41 output stage. An 84 full wave rectifier is used in the power unit. The intermediate frequency is 262 K.C. The diode current establishes a drop across a resistor which is used as additional bias voltage for the R.F. and I.F. tubes giving automatic volume control action. Noise suppression between stations is obtained by the resistor in the cathode circuit of the 75 tube, the drop across which must be overcome before rectification in this tube begins. The manual volume control varies the audio voltage applied to the grid of the 75 tube.

A vibrator interrupts the current through the primary of the power transformer in the power unit. This, together with the turns ratio in this trans-former, results in the high voltage AC being present in the secondary of the transformer. The full wave rectifier tube, filter choke, and filter condensers convert this high voltage AC into high voltage DC for the plate and screen circuits.

Current for the receiver is obtained from the car storage battery. In Fig. 11 is shown the condenser block internal wiring.

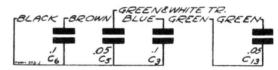

Fig. 11—Condenser Block—Internal Wiring

Voltages at Sockets

In the following chart are given the voltages at the sockets with all the tubes in, all units connected, and the set in operating condition, but with no signal being received. The antenna should be grounded.

A thousand ohm-per-volt meter of 0-250 volt range is required for the plate and screen voltages.

Lower ranges will be necessary for the grid and heater voltages. It is not absolutely necessary to have a high resistance meter for the heater or "A" battery reading.

These voltages will vary with variations in receivers, tubes, test equipment used, and "B" eliminator output voltage.

Type of Tube	Function	Across Heater	Plate to Cathode	Screen to Cathode	Grid to Cathode	Normal Plate MA
78	R. F.	6.1	182	80	3.[1]	7.0
77	1st Det. and Osc.	6.1	178	77	5.[2]	1.3[2]
78	I. F.	6.1	182	80	3.[1]	7.0
75	2nd Det. 1st Audio	6.1	70[3]		1.4[1]	.35
41	Output	6.1	172.5	176.5	12.5[4]	16.0
84	Rect.	6.1	205			17.5 per plate

(1) Cathode to Ground
(2) Subject to Variation
(3) Triode Plate to Cathode
(4) Read Across 400-Ohm Resistor, R13

Vibrator Unit

The vibrator unit is in a small die cast metal box. This box has a paper seal around it. If anything goes wrong with the vibrator unit, return it complete for replacement. Do not dismantle it or order any parts for it.

WELLS - GARDNER COMPANY

Condenser Alignment

Misalignment or mistracking of condensers generally manifests itself in broad tuning and lack of volume at portions or all of the broadcast band. The receivers are all properly aligned at the factory with precision instruments and realignment should not be attempted unless all other possible causes of the faulty operation have first been investigated and unless the service technician has the proper equipment. A signal generator that will provide accurately calibrated signals over the broadcast band and accurately calibrated signals at and around 262 K.C., the intermediate frequency and an output indicating meter are desirable.

First set the signal generator at approximately 262 K.C. Connect the antenna lead from the generator to the control grid of the I.F. 78 tube, through a .05 mfd. condenser. The ground lead of the generator goes to the ground of the receiver. Turn the rotor plates of the tuning condenser completely out and keep the signal weak enough to prevent A.V.C. action. Note from Fig. 10 that the second I.F. transformer is self tuned and cannot be adjusted. Adjust the frequency of the signal generator until the output meter shows maximum output. The intermediate frequency setting of the generator is then correct, although it may be a very small percentage higher or lower than 262 K.C.

Next connect the signal lead from the signal generator to the grid of the 1st detector tube through a .05 mfd. condenser. Then adjust the two intermediate frequency condensers for maximum output. One of the I.F. condenser screws is reached through the hole on the top of the 1st I.F. assembly can. The other I.F. condenser screw is reached from the bottom of the sub-panel through a hole at the bottom of this assembly.

Now set the signal generator for a signal of exactly 1400 K.C. The antenna lead from the generator is, in this instance, connected to the antenna lead of the receiver. Connect the flexible drive shaft to the chassis if it has been disconnected. As explained previously, the dial scale should be at the low frequency end stop when the rotor is completely in mesh. Then turn the station selector knob until the dial scale is at 1400 K.C.

Then adjust the three trimmer condensers on the gang tuning condenser for maximum output, adjusting the oscillator section first.

Next, set the signal generator for a signal of 600 K.C. and adjust the oscillator 600 K.C. trimmer. The adjusting screw for this condenser is reached through a hole in the back wall of the sub-panel.

A non-metallic screwdriver is necessary for this adjustment. Turn the tuning condenser rotor until maximum output is obtained. Then turn the rotor slowly back and forth over this setting, at the same time adjusting the 600 K.C. trimmer screw until the highest output is obtained.

Then set the signal generator again for a signal of 1400 K.C. and check the adjustment of the tuning condenser trimmers at this frequency for maximum output.

Rattle

If rattle is experienced when a signal is being received, it is, in practically all cases, due to mechanical vibration at some point in the chassis. Inspect the chassis and look for a loose tube shield or a loose part at some point which can rattle against another part. When the vibrating part is found, secure it in place in some manner. This can generally be done with a wedge made of a piece of paper, cardboard or wood. Rattle may, in some instances, be due to a loose cover. If this is the case, remove the cover and bend the edge of the chassis box outward between the screw holes so that the cover will fit tightly when it is put on.

If the Receiver Fails to Operate

"A" Fuse — Check the "A" line fuse in the chassis box.

"A" Line Open — See if power is being supplied to the speaker, tube heaters, and "B" eliminator.

"B" Eliminator Not Working — See if the "B" eliminator is in proper working order by checking the high voltage points at the tube plate terminals (see Fig. 10).

Antenna and Lead — See if antenna is properly connected to lead-in wire and antenna lead from set. Be sure antenna system is not grounded at any point.

All Tubes Not Inserted — See if all tubes are inserted as per Fig. 8.

Defective Tubes — Try out a new set of tested tubes.

Grid Caps Not Connected — See if all grid caps are properly connected to top of top grid connection tubes.

Variable Condenser Plates Shorted — Check condenser sections in chassis carefully for foreign particles or rotor stator rubbing.

WELLS - GARDNER COMPANY

Weak Reception

Defective Tubes—Try out a new set of tested tubes and note any difference in performance.

Poor Antenna—To try out the effectiveness of the antenna used, check the volume against the volume when using a straight length of wire about 15' long, run out of the car through one of the windows. If, upon test, the external wire is found to be much superior as far as volume is concerned, the antenna is not satisfactory and will have to be re-vamped or a new one installed. The antenna or lead-in may be too near grounded metal portions of the car frame or body resulting in a high capacity to ground. There may be grounded metal mesh in the car roof. There may be a poor soldered connection between the antenna, lead-in, or antenna lead from the set. The antenna system may be partially grounded at some point.

Antenna Trimmer Not Adjusted—See article "Trying Out the Set and Adjusting."

Car in Shielded Location—If the car is within or near a steel structure, the signals may be weakened by absorption.

Storage Battery Run Down—Check the condition of the battery.

Defective "B" Eliminator—Check "B" voltage at sockets (see voltage chart and Fig. 10).

Misalignment of Variable Tuning Condensers—Instructions for realigning are contained in this manual. Do not, however, attempt realignment unless other causes of low volume have first been investigated.

Wrong Voltages—Check voltages at the sockets (see voltage chart).

Other Causes of Low Volume—Defective speaker, poor battery, antenna, grid cap or other connections, defective A.V.C. system in the receiver, and various opens, grounds and shorts in the receiver assembly.

Distorted Reproduction

Receiver Oscillating—See article on oscillation.

Defective Tubes—Try out a new set of tubes.

Incorrect Voltages—Check the voltages at the sockets (see voltage chart).

Incorrect Tuning—The signal must be carefully tuned in to the clearest and loudest point. It must not be tuned "off resonance."

Defective Speaker—Try out a new one if it is available.

Defective Audio System in the Receiver—Make continuity resistance tests using as a guide Fig. 10.

Signal Transmission—Quality fading in the signal transmission can cause poor tone quality.

Oscillation

Cover of Box—May not be on or if on, may not be sufficiently tightened down.

Off Characteristic Tubes—Tubes whose characteristics vary considerably from the standard may cause oscillation. Try out some new ones.

Open Bypass Condensers—Check the bypass condensers and leads to them for open circuit.

Poor Ground Connections—Check the ground connections in the chassis for poor contact.

Grid Caps and Leads—The grid caps may not be making good contact to the tops of the tubes or the wires of the grid caps may be too close together.

WELLS - GARDNER COMPANY

No. 06W Series
3 UNIT—SUPERHETERODYNE
Automobile Radio
INSTALLATION AND SERVICE MANUAL

Description

The No. 06W Series Auto Radio Receivers are made up in three units: the chassis unit, speaker—"B" eliminator unit and control unit. The control unit is mounted on the steering column, while the speaker—"B" eliminator unit and chassis are mounted on the dash. Current to operate the chassis and "B" eliminator is obtained from the automobile storage battery. Two flexible shafts mechanically connect the control unit to the chassis. One of these is for the volume control and switch, while the other is for the tuning mechanism. A roof antenna is recommended.

In this manual are covered detailed instructions for the installation of each part and information for completing and maintaining the installation. Data for servicing the receiver is also included, should the necessity for such procedure arise. The following tools are required: portable electric drill, screw drivers, pliers, a heavy soldering iron, hack saw, files, small wrenches, and diagonal cutters.

Before making the installation it is suggested that this manual be completely read.

Mounting the Control Unit

The control unit is mounted on the steering column under the steering wheel as shown in Figs. 1 and 2. A clamp is used to hold it in position.

The outer portion of the clamp is screwed to the inner portion by means of the four 8-32x⅜" fillister head screws supplied with the receiver.

Two rubber strips are provided, one ⅛" thick and the other 1/16" thick. These are wrapped around the steering column under the clamp. Either or both of these strips may be used, depending on the thickness of the column. Wrap the rubber strips around the column in such a way as to allow the set screws which hold the clamp in position to pass through. When the clamp is in place, take the two

8-32 headless cup point set screws and screw them down on the steering column through the holes in the clamp.

The control unit is generally about 4" below the wheel, but this will vary with individual cases. The length of the drive shaft and interference with driver's legs will also govern the location of the control unit.

There are two screws which hold the inside portion of the clamp to the bracket on the box. By loosening these two screws, the box can be swung around if such a position is handier from the standpoint of the person operating the set. Instructions for attaching the pilot lamp are contained in the article "Completing the Wiring Connections."

Mounting the Chassis

The chassis is mounted in back of the dash as shown in Figs. 1 and 2. It should be mounted in such a way that the flexible drive shafts to the control unit will be in as straight a line as possible. The chassis is mounted with the anchor bushing into which the flexible drive shafts go, facing the control unit. In the illustrations mentioned above, the

chassis is on the right side of the dash which is a good location from the standpoint of flexible drive shaft arrangement. *Before mounting the chassis read the section on "Attaching the Flexible Drive Shafts."*

The chassis is secured to the dash by means of the dash mounting plate (see Fig. 3). First drill the

WELLS - GARDNER COMPANY

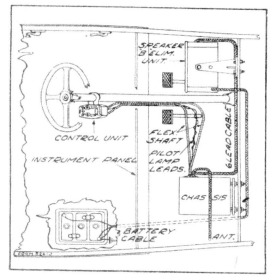

Fig. 1—General Installation—Top View

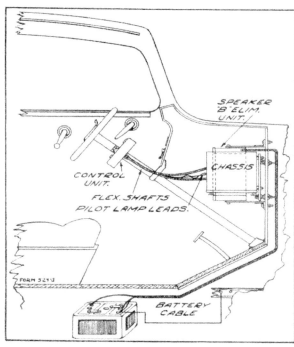

Fig. 2—General Installation—Side View

three mounting holes required for the dash mounting plate. The location and size of these holes is shown in Fig. 3. A template for drilling these holes is supplied with the set. Three 4" square head mounting bolts are supplied. Take two of these, which will be used for the upper part of the mounting plate and screw on nut "A" (see Fig. 4). The nut should be just far enough away from the head of the bolt to permit the bracket of the mounting plate to slip down as shown in the illustration. Then put on nut "B" and the washer, after which the two bolts can be put through the dash, with the shanks extending into the engine compartment, as shown in Fig. 4. A washer, lockwasher, and nut are then put on these bolts from the front of the dash to hold them in place.

The distance "X" between nuts "A" and "B" determines how far out the chassis is mounted from the dash. When there is a lot of apparatus in back of the dash, such as wires, tubing, etc., the chassis will have to set out far enough to clear it. However, in most cars, there is no interfering apparatus and therefore the distance "X" will be zero.

Then put a washer on the third mounting bolt and put this bolt through the lower mounting hole with the head on the engine side of the dash, as shown in the illustration. Put on a washer, lockwasher, and nut "D" and tighten it up. Then put on nut "E" with a washer as shown. Nut "E" should be screwed down until it is about $\frac{1}{4}$" from nut "D," when distance "X," as explained above, is zero.

Next, secure the dash mounting plate to the chassis box by means of the four chassis mounting screws.

Note that the broad or narrow face of the chassis box can be secured to the dash mounting plate. Use whichever side will be best from the standpoint of attachment of the flexible drive shafts.

All the tubes should be in the sockets, the antenna trimmer adjusted (as explained later) and the flexible drive shafts connected before the chassis is permanently installed. Complete information on the latter procedure is contained in the article on attaching the flexible drive shafts.

The four mounting screws pass through the four slots in the mounting plate (Fig. 3). After they are in place and tight, the dash mounting plate with chassis attached is slipped over the three mounting bolts. The two upper brackets on the plate slip down in back of nut "A" as shown in Fig. 4 and the slot at the bottom of the plate slips over the shank of the lower mounting bolt in back of nut "E." The plate will then hang with the bottom farther away from the dash than the top. A washer, lockwasher, and nut "F" are then put on the lower mounting bolt. Nut "F" is screwed on until the mounting plate is tight up against the washer in back of nut "E." In this position, the bracket at the top of the mounting plate should butt up against nut "A" and be tight. Also the mounting plate will be approximately parallel with the dash.

WELLS - GARDNER COMPANY

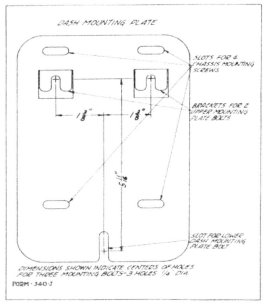

Fig. 3—Dash Mounting Plate

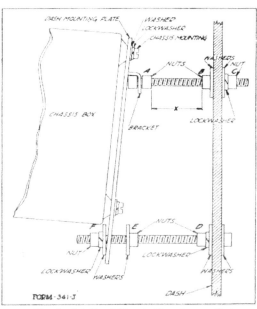

Fig. 4—Details of Chassis Mounting on Dash

Mounting the Speaker-"B" Eliminator

The speaker-"B" eliminator is mounted on the back of the dash by means of two brackets, as shown in Fig. 5. Usually the space available will govern the location of the speaker and position of it on the mounting brackets. However, the matter of acoustics should be given careful consideration. One of the most desirable positions from the standpoint of

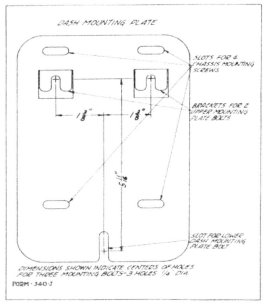

Fig. 5—Method of Mounting Speaker

acoustics is that shown by the solid lines in Fig. 5 (A). In this position the sound waves travel in the most direct lines toward the listener. After the

speaker is mounted and regardless of the position of the brackets, loosen the bracket bolts and turn it to several positions in order to get the best one from the standpoint of tone quality.

Other considerations governing the location of the speaker are the cables and the tone control. The speaker should be so mounted that the two shielded cables, one to the storage battery and one to the chassis, will be long enough and can be most conveniently brought over. The tone control knob on the speaker box should be preferably on the bottom, so that it can be reached easily.

After the position of the speaker is decided on, drill the four ₅/₁₆" holes required for the bracket mounting bolts. A template for these holes is supplied with the receiver. The holes are arranged in a rectangle. The centers of the holes, the small dimension are 2⅜" apart and the long dimension 10" apart. In Fig. 5 is shown how the brackets can be mounted horizontally (A) or vertically (B), and the different positions in which the speaker itself can be placed. There are two holes in each bracket as shown in Fig. 5 (C) which determine the distance of the speaker box from the dash. The grilled portion of the box at the front should face the listener.

WELLS-GARDNER COMPANY

Attaching the Flexible Drive Shafts

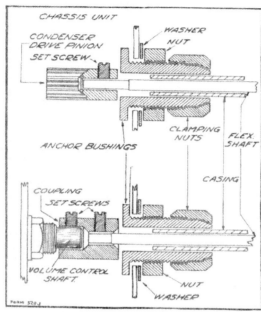

Fig. 6—Details of Flexible Drive Shaft Connections

After the control unit and chassis are in position, the flexible drive shafts may be attached. Two 34" shafts are supplied, unless otherwise specified. These shafts may also be had in 14", 20", and 45" lengths.

The flexible drive shafts should be put on with a minimum amount of bending. In general, one large radius 90° bend is all that is necessary.

The 34" shafts supplied with the receiver may be cut to a shorter length if necessary. The shaft (inside portion) should first be brazed at the point to be cut. It should then be cut with a three-corner file or edge of a grinding wheel. *Do not use a hack saw.* The casing which is 1½" shorter must be cut to correspond. This should be tinned or brazed first at the point to be cut and may then be cut with a hack saw.

After the length and position of the shafts is decided on, remove the chassis and mounting plate from the mounting bolts. As the shafts are already secured at the control unit, it is necessary only to

If the flexible shaft is cut as mentioned above, file it down in one place a slight amount to provide a flat surface for the set screw.

A roof antenna is recommended, as by far the best results will be obtained. A large percentage of cars at the present time come equipped by the factory with built-in roof antennas. In those cars which do not have an antenna, one will have to be put in.

secure them at the chassis end. Before attaching the shafts, see if the set is in working order. Put the 8-prong socket in place on the chassis and operate the set with the cover off.

In Fig. 6 is shown a cross-sectional view of the flexible drive shaft connections at the chassis end. First put the tube cover plate on the chassis box temporarily with two screws. This is the large plate held in position ordinarily by means of five screws. Then center the volume control anchor bushing on this plate. To do this, loosen the nut which holds this bushing in place (see Fig. 6). Center the bushing by eye so that the center of it is in a line with the center of the volume control coupling. Then tighten the nut down.

Next, take the tube cover plate off. Extend the volume control flexible shaft and casing several inches through the hole in the anchor bushing of the tube cover plate so that the plate will be on the casing and out of the way. Turn the volume control coupling counter-clockwise until the switch is snapped to the off position. Lock the receiver on the control unit and turn the volume control knob counter-clockwise until it is in the locked position. Then loosen both set screws in the volume control coupling and insert the flexible shaft in the coupling (see Fig. 6). Tighten the outer set screw first on one of the four flat faces of the flexible shaft and then tighten the inner set screw. Then again temporarily hang the chassis on the mounting bolts. Next, check the operation of the switch, volume control and lock. The switch should be off when the volume control knob is in the locked position. It may be necessary to loosen the inner set screw and do a slight amount of adjusting until the proper setting is obtained.

Next, slide the tube cover plate into position and fasten it in place by means of the five screws. Then tighten the clamping nut on the volume control shaft casing but *do not tighten this nut excessively.*

To attach the tuning condenser flexible shaft, first center the anchor bushing by eye as was explained above. Then extend the tuning condenser flexible shaft into the hole at the center of the tuning condenser drive pinion. With the rotor plates completely in mesh, turn the dial gear in the control unit until it is at the low frequency end stop. The set screw may then be tightened and the clamping nut secured on the casing as was explained above. In some instances, it may be necessary to loosen the set screw of the large gear on the tuning condenser rotor shaft and adjust the setting of this gear in order to get an accurate calibration.

Antenna

First determine if the top has a grounded chicken wire mesh. To do this, use a continuity meter. By means of a wire, attach a darning needle to one of the prods. Poke the darning needle into the roof material and turn it around until it comes in con-

WELLS-GARDNER COMPANY

tact with the chicken wire. Then ground the other prod and if the continuity meter shows a complete circuit, the chicken wire mesh is grounded. In a case of this kind, it will be necessary to get inside of the roof and it is advisable to employ the services of an auto "top man" or an upholsterer.

It will be necessary to remove the top material and cut away the chicken wire from the side supports until it is at least 3" away from the ground at any point. It should also be at least 3" away from the dome light and the dome light wiring. The chicken wire may then be laced to the points from which it was cut with a heavy, waxed cord. The chicken wire will then make a satisfactory antenna, or a copper screen may be used.

If the chicken wire is not grounded, it may be used as the antenna by taking down the roof material at one corner and soldering the lead-in wire to it. If it

Pilot Lamp

The pilot lamp cable is 4 feet long and is attached to the 8-prong socket. At the end of the cable is the pilot lamp socket and spring clip. After the control unit and chassis are mounted, remove the cover of the control unit by taking off the two knobs, the key entry nut and the cover screw. Bring the pilot lamp

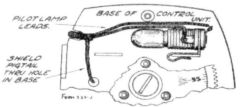

Fig. 7—Pilot Lamp Attachment

cable through the notch at the side of the back of the unit. Then, clip the pilot lamp socket clip over the right hand bracket as shown in Fig. 7, with the two leads going over the top of the lamp as illustrated. It is not necessary to remove the dial gear. There is a "pigtail" on the end of the shield of the pilot lamp cable. Pull this "pigtail" through the hole beneath the slot, as shown in the illustration. Then insert the round head 3/8" 8-32 screw through this hole with the head on the outside of the box and secure it in place with the lockwasher and nut provided. This holds the "pigtail" in position and

is not desired to take down the roof material a piece of copper screening can be tacked to the roof on the inside of the car. At least six square feet should be used. Keep it at least 3" away from any grounded metal parts on all sides. After the screen is in place, it can be covered over with cloth which matches the roof material. Solder the lead-in wire to the screen and bring it down the front corner post nearest to the set.

Another, and a very simple way in which an antenna can be secured to the inside of the car roof is to use a car-roof antenna which is made up especially for this purpose. This antenna consists of copper strips laid back and forth between two pieces of cardboard and the center being covered over with material which matches the roof material. It can be had in several colors and is tacked in place on the inside of the car roof in a few minutes.

grounds it. Cut off the excess length of "pigtail." Double up the pilot lamp leads if too long—do not cut them.

Antenna Cable

Bring the antenna cable of the receiver in the most direct manner possible to the lead-in from the antenna and connect it to the latter. Keep it as high as possible and as far away from any car wiring as possible. Care should be taken not to have the antenna wire come in contact with the shield wires. Ground the shield of the antenna cable at the antenna end.

Battery Cable and Six Lead Cable

The battery cable should be brought over to the storage battery in the most convenient manner possible. In Figs. 1 and 2 it is shown passing through a hole in the dash, thence down and under the floor board to the battery. In other installations, it may be more convenient to bring this cable down in back of one of the side pads and thence to the battery. The lug on the lead marked "positive" is connected to the positive side of the battery and the lug on the negatively marked lead is connected to the negative side of the battery. Ground the pigtail of the shield by screwing the No. 6 Parker Kalon screw through the end of the pigtail and through the hole in the lug which is grounded.

The six-lead cable between the chassis and the speaker—"B" eliminator is usually brought over along the dash in the most convenient manner possible.

Trying Out the Set and Adjusting

After the wiring has all been completed and before the chassis is permanently installed, try out the set and adjust the antenna trimmer condenser. The

location of the tubes is shown in Fig. 8. Do not start the engine of the car yet.

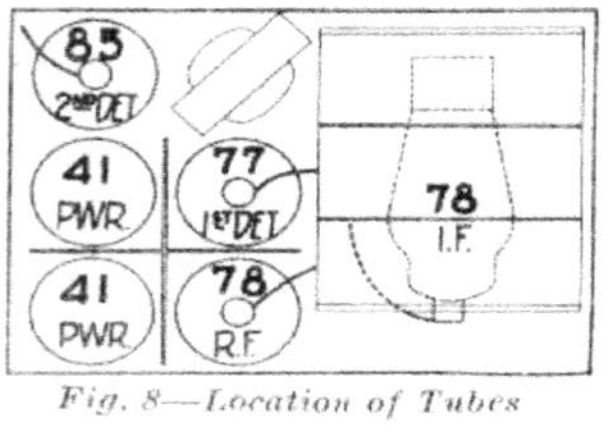

To adjust the antenna trimmer, tune in a weak

Fig. 8—Location of Tubes

signal between 1200 and 1400 KC with the volume control about three-quarters on. On one end of the chassis box is a small metal plate. Remove the two screws which hold this plate in place. Directly under the hole in the chassis box is the antenna trimmer condenser screw. Turn this adjusting screw up or down until maximum output is obtained.

If the receiver fails to operate, check the items as given under the article by that name.

WELLS-GARDNER COMPANY

current establishes a drop across a resistor which is used as additional bias voltage for the R.F. and I.F. tubes giving automatic volume control action. Noise suppression between stations is obtained by the resistor in the cathode circuit of the 85 tube, the drop across which must be overcome before rectification on this tube begins. The manual volume control varies the audio voltage applied to the grid of the 85 tube.

The "B" eliminator and speaker are in one box. A vibrator interrupts the current through the primary of the transformer in the "B" eliminator. Another vibrator in the secondary circuit operating at the same frequency acts as a rectifier. The on-off relay in the "B" eliminator closes the primary circuit when the set switch is turned on. The load relay provides a load current for the secondary circuit if the "B" line is drawing less than normal current.

Current for the speaker field, tube heaters, and "B" eliminator is obtained from the car storage battery. In Fig. 11 is shown the condenser block internal wiring.

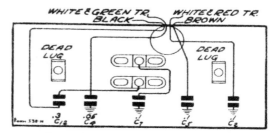

Fig. 11—Condenser Block—Internal Wiring

Condenser Alignment

Misalignment or mistracking of condensers generally manifests itself in broad tuning and lack of volume at portions or all of the broadcast band. The receivers are all properly aligned at the factory with precision instruments and realignment should not be attempted unless all other possible causes of the faulty operation have first been investigated and unless the service technician has the proper equip-

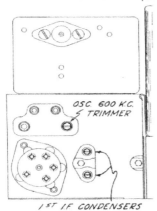

Fig. 12—Location of Trimmers

ment. A signal generator that will provide accurately calibrated signals over the broadcast band and accurately calibrated signals at and around 262 K.C., the intermediate frequency and an output indicating meter are desirable.

First set the signal generator at approximately 262 K.C. Connect the antenna lead from the generator to the control grid of the I.F. 78 tube, through a .05 mfd. condenser. The ground lead of the generator goes to the ground of the receiver. Turn the rotor plates of the tuning condenser completely out

and keep the signal weak enough to prevent A.V.C. action. Note from Fig. 10 that the second I.F. transformer is self tuned and cannot be adjusted. Adjust the frequency of the signal generator until the output meter shows maximum output. The intermediate frequency setting of the generator is then correct, although it may be a very small percentage higher or lower than 262 K.C.

Next connect the signal lead from the signal generator to the grid of the 1st detector tube through a .05 mfd. condenser. Then adjust the two intermediate frequency condensers for maximum output. The location of the adjusting screws for these condensers is shown in Fig. 12.

Now set the signal generator for a signal of exactly 1400 K.C. The antenna lead from the generator is, in this instance, connected to the antenna lead of the receiver. Connect the flexible drive shaft to the chassis if it has been disconnected. As explained previously, the dial scale should be at the low frequency end stop when the rotor is completely in mesh. Then turn the station selector knob until the dial scale is at 1400 K.C.

Then adjust the three trimmer condensers on the gang tuning condenser for maximum output, adjusting the oscillator section first (section farthest from drive gear).

Next set the signal generator for a signal of 600 K.C. and adjust the oscillator 600 K.C. trimmer. The location of this condenser is shown in Fig. 12.

A non-metallic screwdriver is necessary for this adjustment. Turn the tuning condenser rotor until maximum output is obtained. Then turn the rotor slowly back and forth over this setting, at the same time adjusting the 600 K.C. trimmer screw until the highest output is obtained.

Then set the signal generator again for a signal of 1400 K.C. and check the adjustment of the tuning condenser trimmers at this frequency for maximum output.

Circuit

The circuit consists of an antenna stage, a 78 R.F. stage, a 77 1st detector-oscillator stage, a 78 I.F. stage, an 85 duo-diode-triode tube which functions

as a diode 2nd detector and triode 1st audio stage, and two 41 tubes in a semi-Class "B" output stage. The intermediate frequency is 262 K.C. The diode

WELLS-GARDNER COMPANY

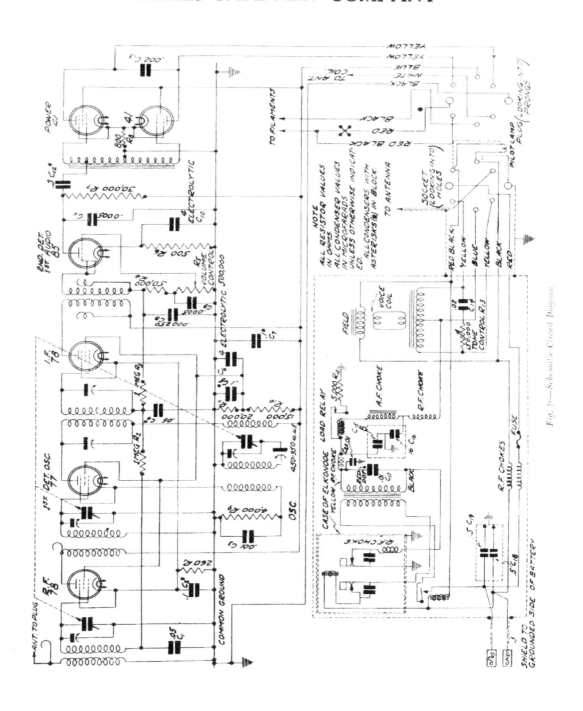

Fig. 10—Schematic Circuit Diagram

WELLS-GARDNER COMPANY

Voltages at Sockets

In the following chart are given the voltages at the sockets with all the tubes in, all units connected, and the set in operating condition, but with no signal being received. The antenna should be grounded.

A thousand ohm-per-volt meter of 0-250 field range is required for the plate and screen voltages.

Lower ranges will be necessary for the grid and heater voltages. It is not absolutely necessary to have a high resistance meter for the heater or "A" battery reading.

These voltages will vary with variations in receivers, tubes, test equipment used, and "B" eliminator output voltage.

Type of Tube	Function	Across Heater	Plate to Cathode	Screen to Cathode	Grid to Cathode	Normal Plate MA
78	R. F.	6.1	182	80	3.(1)	7.0
77	1st Det. and Osc.	6.1	178	77	5.(2)	1.3(2)
78	I. F.	6.1	182	80	3.(1)	7.0
85	2nd Det. 1st Audio	6.1	70(3)		1.8(1)	3.5
41	Output	6.1	162	168.5	17.	11.

NOTE:—Above voltages are at 185 volts input from "B" Eliminator.
(1) Cathode to Ground
(2) Subject to Variation
(3) Triode Plate to Cathode

Elkonode Unit

The Elkonode Unit is the small metal encased box in the "B" eliminator which contains the two vibrators. If anything goes wrong with this unit, return it complete for replacement. Do not dismantle it or order any parts for it.

If the Receiver Fails to Operate

"A" Fuse—Check the "A" line fuse in the speaker box.

"A" Line Open—See if power is being supplied to the speaker, tube heaters, and "B" eliminator.

"B" Eliminator Not Working—See if the "B" eliminator is in proper working order by checking the high voltage points at the speaker-terminal strip and at the tube plate terminals (see Fig. 10).

Antenna and Lead—See if antenna is properly connected to lead-in wire and antenna lead from set. Be sure antenna system is not grounded at any point.

All Tubes Not Inserted—See if all tubes are inserted as per Fig. 8.

Grid Caps Not Connected—See if all grid caps are properly connected to top of top grid connection tubes.

Variable Condenser Plates Shorted—Check condenser sections in chassis carefully for foreign particles or rotor stator rubbing.

Reversed Storage Battery Connections—Check storage battery connections for correctness.

Weak Reception

Defective Tubes—Try out a new set of tested tubes and note any difference in performance.

Poor Antenna—To try out the effectiveness of the antenna used, check the volume against the volume when using a straight length of wire about 15' long, run out of the car through one of the windows. If, upon test, the external wire is found to be much superior as far as volume is concerned, the antenna is not satisfactory and will have to be re-vamped or a new one installed. The antenna or lead-in may be too near grounded metal portions of the car frame or body resulting in a high capacity to ground. There may be grounded metal mesh in the car roof. There may be a poor soldered connection between the antenna, lead-in, or antenna lead from the set. The antenna system may be partially grounded at some point.

Antenna Trimmer Not Adjusted—See article "Trying Out the Set and Adjusting."

WELLS-GARDNER COMPANY

No. V6Z2 Series

"AUTO-COMPACT" WITH "MONO"
STEERING COLUMN CONTROL

Automobile Radio

INSTALLATION AND SERVICE MANUAL

Description

The No. V6Z2 series receivers are made up in two units: the chassis unit containing the radio chassis, speaker and "B" eliminator; and the control unit. The chassis is mounted on the dash by means of two brackets, while the control unit is mounted on the steering column by means of a clamp. Two flexible shafts mechanically connect the control unit to the chassis. One of these is for the volume control and switch, while the other is for the tuning mechanism. A roof antenna is recommended. Current to operate the receiver is obtained from the car storage battery.

In this manual are covered detailed instructions for the installation of each part and information for completing and maintaining the installation. Data for servicing the receiver is also included, should the necessity for such procedure arise. The following tools are required: portable electric drill, screw drivers, pliers, a heavy soldering iron, hack saw, files, small wrenches, and cutters.

Before making the installation it is suggested that this manual be completely read.

Mounting the Control Unit

The control unit is mounted on the steering column under the steering wheel as shown in Figs. 2 and 3. A clamp is used to hold it in position.

The outer portion of the clamp is screwed to the inner portion by means of the four 8-32x¾" fillister head screws supplied with the receiver. See Fig. 1.

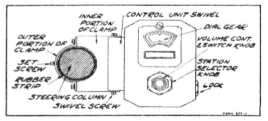

Fig. 1—Method of Mounting Control Unit

Two rubber strips are provided, one ⅛" thick and the other ₁₆" thick. These are wrapped around the steering column under the clamp. Either or both of these strips may be used, depending on the thickness of the column. Wrap the rubber strips around the column in such a way as to allow the set screws which hold the clamp in position to pass through. When the clamp is in place, take the two 8-32 headless cup point set screws and screw them down on the steering column through the tapped holes in the clamp.

The control unit is generally about 4" below the wheel, but this will vary with individual cases. The length of the drive shaft and interference with driver's legs will also govern the location of the control unit.

There are two screws which hold the inside portion of the clamp to the control unit swivel. By loosening these two screws, the box can be swung around if such a position is handier from the standpoint of the person operating the set. Instructions for attaching the pilot lamp lead are contained in the article "Completing the Wiring Connections."

Mounting the Chassis

The chassis is mounted on the dash by means of two brackets as shown in Fig. 2. Two mounting screws are used to secure each bracket to the end of the chassis box. Six embossings with inset nuts are provided on each end of the chassis box. Any two of these may be used for the bracket screws, thus pro-

WELLS-GARDNER COMPANY

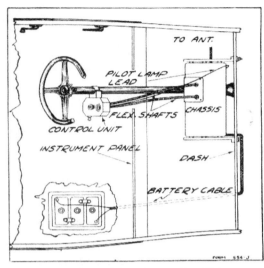

Fig. 2—General Installation—Top View

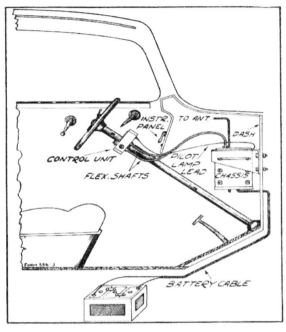

Fig. 3—General Installation—Side View

viding great flexibility in mounting.

Each nut has a mounting screw in it and if any of these are in the way of the mounting bracket, they can be taken out.

The chassis should be mounted with the speaker grill facing toward the driver. In this position, the anchor bushings in which the flexible drive shafts are placed will come out of the top.

The location of the chassis will very often depend on the space available. To the left of the center, as shown in Fig. 2, is a good location. The chassis should be mounted in such a way that the flexible drive shafts to the control unit will be in as straight a line as possible or with large radius bends. *In general, it will be advisable to consider the possibility of a car heater installation at the right side of the dash (facing forward).* In practically every case no difficulty will be experienced in mounting the heater and chassis on the dash.

The possibility of interference with the legs of the driver or passenger in the front seat and the possibility of interference with the controls of the car should also be considered before the location of the chassis is definitely decided on.

Before mounting the chassis read the section on "Attaching the Flexible Drive Shafts."

When the location is decided on, drill the four mounting holes required. The location and size of these holes is shown in Fig. 4. A template for drilling these holes is supplied with the receiver. Four ¼" mounting bolts, four washers, four lockwashers, and four nuts are provided. The mounting bolt is put through the bracket and dash with the shank

Fig. 4—Mounting Hole Location

extending into the engine compartment. A washer, the lockwasher and nut are then put on. Mount the brackets permanently, but do not mount the chassis permanently until all connections are completed, the tubes are all inserted, the receiver tried out, and the antenna trimmer adjusted (explained later).

Attaching the Flexible Drive Shafts

After the control unit and chassis are in position, the flexible drive shafts may be attached. Two 34" shafts are supplied, unless otherwise specified. These shafts may also be had in 14", 20", and 45" lengths.

The flexible drive shafts should always be installed with a minimum amount of bending. Always keep the radius of the bend as large as possible.

The 34" shafts supplied with the receiver may be cut to a shorter length if necessary. The shaft (inside portion) should first be brazed at the point to be cut. It should then be cut with a three-corner file or edge of a grinding wheel. *Do not use a hack saw.* After the shaft is cut, file it down in one place a slight amount to provide a flat surface for the set

WELLS-GARDNER COMPANY

screw. The casing which is 1½″ shorter must be cut to correspond. This should be tinned or brazed first at the point to be cut and may then be cut with a hack saw.

After the length and position of the shafts is decided on they may be secured to the chassis. The shafts are already secured at the control unit. It is advisable to attach the flexible shafts with the chassis on the mounting brackets, but if the chassis is accessible, it may be removed from the brackets. Keep it as close to its regular position as possible so that the flexible shaft will not turn after the chassis is replaced on the brackets. In general, it may be moved up or down, but should not be moved sideways or be turned. Just over the speaker grill on the chassis box will be seen an angle plate. Remove this plate. Before proceeding further with attachment of the shafts see if the receiver is in working order by operating it with the cover off and necessary connections completed, as explained further in this manual.

In Fig. 5 is shown a cross-sectional view of the flexible drive shaft connections at the chassis end. First put the angle plate on the chassis box temporarily with two screws. Then center the volume control anchor bushing on this plate. To do this, loosen the nut which holds this bushing in place (see Fig. 5). Center the bushing by eye so that the center of it is in a line with the center of the volume control coupling. Then tighten the nut down.

Next, take the angle plate off. Extend the volume control flexible shaft and casing several inches through the hole in the anchor bushing of the angle plate so that the plate will be on the casing and out of the way. Turn the volume control coupling counter-clockwise until the switch is snapped to the off position. Lock the receiver on the control unit and turn the volume control knob counter-clockwise until it is in the locked position. Then loosen both set screws in the volume control coupling and insert the flexible shaft in the coupling (see Fig. 5). Tighten the outer set screw first on one of the *four flat faces* of the flexible shaft and then tighten the inner set screw. For purposes of illustration, the set screws in Fig. 5 are shown extending sideways in the coupling, but should actually extend towards the box opening in order to get at them. Then temporarily place the chassis on the mounting brackets if it has been taken off and check the operation of the switch, volume control, and lock. The switch should be off when the volume control knob is in the locked position. It may be necessary to loosen the inner set screw and do a slight amount of adjusting until the proper setting is obtained.

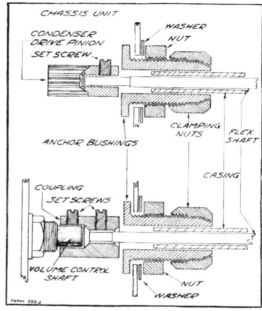

Fig. 5—Details of Flexible Drive Shaft Connections

To attach the tuning condenser flexible shaft, first center the anchor bushing by eye as was explained above. Then extend the tuning condenser flexible shaft into the hole at the center of the tuning condenser drive pinion. Turn the large gear on the tuning condenser rotor shaft until the rotor plates are completely in mesh. Then turn the station selector knob on the control unit until the dial gear is at the low frequency end stop. The set screw in the drive pinion should then be tightened down on one of the four flat faces of the shaft.

The operation of this control should also be tried out after the shaft is in place. In order to get accurate calibration it may be necessary in some instances to loosen the set screw of the large gear on the tuning condenser rotor shaft and adjust the setting of this gear.

Next, slide the angle plate into position and fasten it in place by means of the four screws. Then tighten down the clamping nuts on the two flexible shaft casings, *but do not tighten these nuts excessively.*

Antenna

A roof antenna is recommended, as by far the best results will be obtained. A large percentage of cars at the present time come equipped by the factory with built-in roof antennas. In those cars which do not have an antenna, one will have to be put in.

First determine if the top has a grounded chicken wire mesh. To do this, use a continuity meter. By means of a wire, attach a darning needle to one of the prods. Poke the darning needle into the roof material and turn it around until it comes in con-

WELLS-GARDNER COMPANY

tact with the chicken wire. Then ground the other prod and if the continuity meter shows a complete circuit, the chicken wire mesh is grounded. In a case of this kind, it will be necessary to get inside of the roof and it is advisable to employ the services of an auto "top man" or an upholsterer.

It will be necessary to remove the top material and cut away the chicken wire from the side supports until it is at least 3" away from ground at any point. It should also be at least 3" away from the dome light and the dome light wiring. The chicken wire may then be laced to the points from which it was cut with a heavy, waxed cord. The chicken wire will then make a satisfactory antenna, or a copper screen may be used.

If the chicken wire is not grounded, it may be used as the antenna by taking down the roof material at one corner and soldering the lead-in wire to it. If it is not desired to take down the roof material a piece of copper screening can be tacked to the roof on the inside of the car. At least six square feet should be used. Keep it at least 3" away from any grounded metal parts on all sides. After the screen is in place, it can be covered over with cloth which matches the roof material. Solder the lead-in wire to the screen and bring it down the front corner post nearest to the set.

Another, and a very simple way in which an antenna can be secured to the inside of the car roof is to use one of the car-roof antennas which are now being made up especially for this purpose. There is one type of antenna which consists of copper strips laid back and forth between two pieces of cardboard. The cardboard is then covered over with material which matches the roof material. This antenna can be had in several colors and is tacked in place on the inside of the car roof in a few minutes.

Completing the Wiring Connections

Pilot Lamp

The pilot lamp lead is in a shielded cable which extends out from the control unit box. On the rear wall of the chassis, near one of the ends, will be seen a tip jack. Insert the tip on the end of the pilot lamp lead into this jack. There is also a pigtail or shield extension at the end of this lead. Ground this pigtail with one of the angle plate screws (see Fig. 6). Double up the pilot lamp lead if it is too long—*Do not cut this lead.*

Antenna Cable

Bring the antenna cable of the receiver in the most direct manner possible to the lead-in from the antenna and connect it to the latter. Keep it as high as possible and as far away from any car wiring as possible. Care should be taken not to have the antenna wire come in contact with the shield wires. Ground the pigtail of the antenna cable shield at the antenna end. The pigtail of this shield at the chassis end is grounded under one of the chassis mounting screws.

In some cases the shielded antenna lead from the receiver is not long enough to reach to the column at which the antenna lead-in comes down. In a case of this kind, cover the exposed portion of the lead-in wire with braided shield from the point where it leaves the column to the point of connection to the antenna lead of the receiver. Connect the two wires together and connect the two shields together, care being taken that no strand of the shield touches the antenna wire.

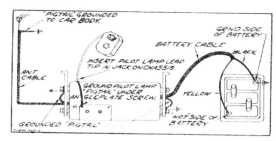

Fig. 6.—External Wiring Connections

Battery Cable

The battery cable should be brought over to the storage battery in the most convenient manner possible. In Figs. 2 and 3 it is shown passing through a hole in the dash, thence down and under the floor board to the battery. In other installations, it may be more convenient to bring this cable down in back of one of the side pads and thence to the battery. The lug on the yellow lead of this cable is connected to the "Hot" or ungrounded side of the battery (the "Hot" or ungrounded side may be positive or negative, depending on the make of car). The lug on the black lead is connected to the grounded side of the battery. The pigtail of the shield of this cable at the chassis end should be grounded under one of the chassis mounting screws.

Trying Out the Set and Adjusting

After the wiring has all been completed and before the chassis is permanently installed, try out the set and adjust the antenna trimmer. The location of the tubes is shown in Fig. 7. To adjust the antenna trimmer, tune in a weak signal between 1200

Fig. 7.—Location of Tubes

and 1400 K.C. with the volume control about three-fourths on. On one end of the chassis box is a small metal plate. Remove this plate. Directly under the hole in the chassis box is the antenna trimmer condenser screw. Turn this adjusting screw up or down until maximum output is obtained.

WELLS-GARDNER COMPANY

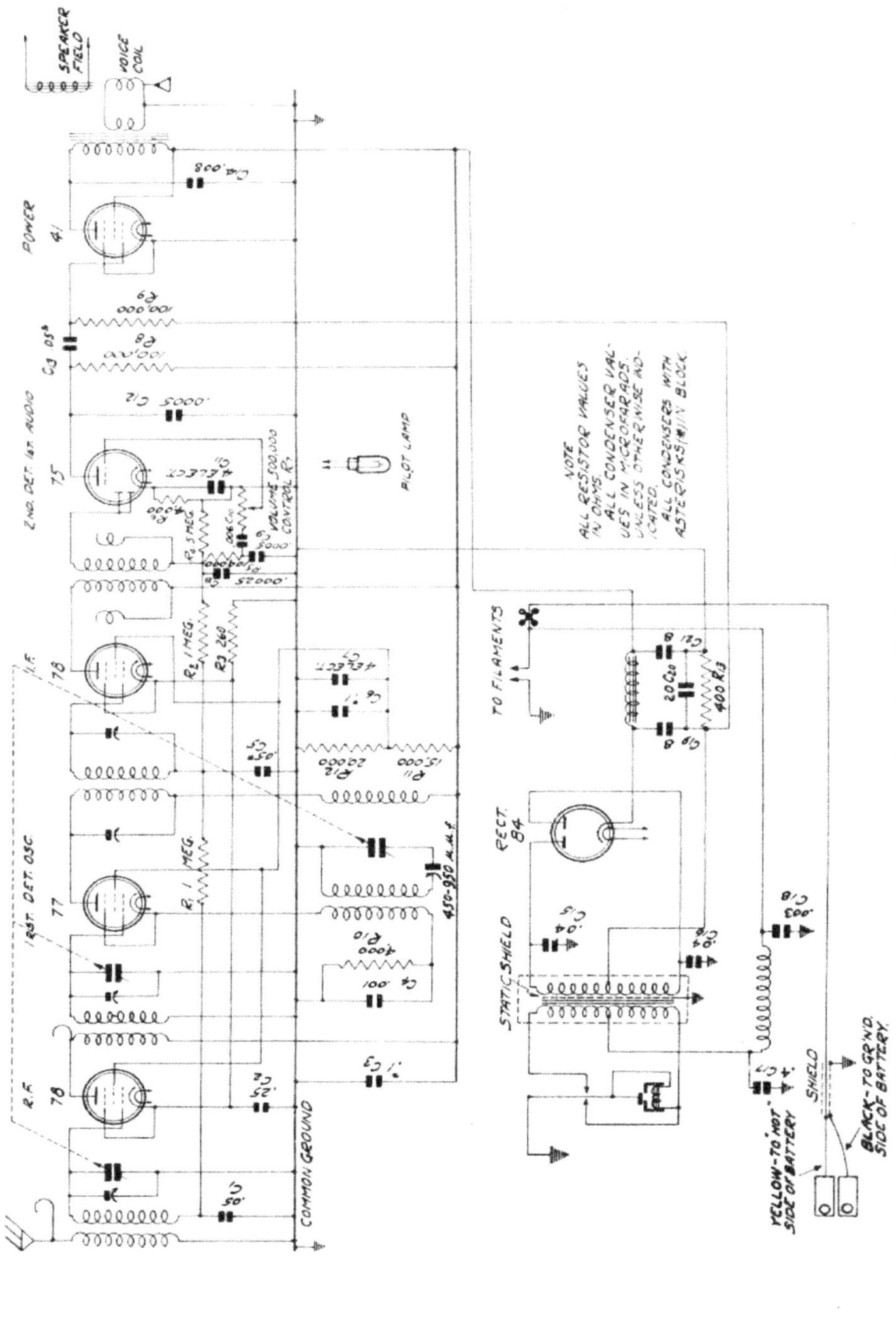

Fig. 9—Schematic Circuit Diagram

WELLS-GARDNER COMPANY

Ground all cables and tubing which pass through the dash, such as oil lines, gas lines, etc. Ground to the dash or at the nearest convenient point on the frame with a good short ground connection. Use braided shield for this purpose.

If the chassis and coil are both in back of the dash (under the cowl), take off the coil and mount it on the front of the dash (in the engine compartment). If the coil cannot be moved, place a copper can over it and ground the can at the coil mounting.

In some cases, the high and low tension leads between the coil and distributor are run close together. In some cars they are in the same conduit. If this is the case, remove the low tension lead from this conduit. In any event, keep the high and low tension leads as far apart from each other as possible. Shield and ground the shield of the high tension lead, if separating the two leads is not sufficient. Then try also shielding the low tension lead.

The motor must, in every case, be well grounded to the frame of the car. If it is not, use a very heavy braided lead for this purpose, similar to a storage battery ground lead.

Noise, on occasion, may be due to weak pickup caused by the automobile being in a shielded location. The action of the automatic volume control, due to the low pickup, causes the set to operate at its maximum sensitivity, thereby increasing noisy reception, due both to external pickup and internal conditions.

Noisy operation is also caused in some instances by loose parts in the car body or frame. These loose parts rubbing together affect the grounding and cause noises, due to the rubbing or wiping action. Tightening up the frame and body at all points and in some cases, the use of a copper jumper will eliminate noise of this nature.

Care and Maintenance

Advancing Generator Charging Rate

The installation of the automobile radio imposes an additional drain on the car storage battery. This can be compensated for by advancing the charging rate of the car generator. Check the state of charge of the storage battery about a week after the installation of the automobile radio is made and adjust the charging rate accordingly.

Tubes

The type of tubes used and location of these tubes in the chassis are shown in Fig. 7. These tubes are of a sturdy, rugged construction designed especially for an auto receiver. Most of them, under normal usage, will last for many months and in some cases, years. Some of them, however, may become faulty after a few months of operation.

For that reason, it is advisable to secure a new set of tested tubes at intervals of three to six months and have these inserted in the receiver one at a time, noting any difference in performance.

Pilot Lamp

The pilot lamp is located in the control unit. A 6-8 volt miniature base lamp is used. To replace the lamp, first turn the receiver off. Remove the two control knobs and the key entry nut. Then take out the screw holding the control box cover in place after which the cover can be taken off. The pilot lamp socket is secured to a spring clip which is on a bracket in the control unit. Push this clip and socket over far enough to get at the lamp, after which the bulb can be replaced and the control unit reassembled.

Fuse

A 10 amp. automobile fuse is used for the "A" line. This fuse is mounted on a block on the power transformer in the chassis. To change the fuse, it will be necessary to remove the cover of the chassis box.

Electrical Condition of Car

Dirty spark plugs, incorrect spacing of distributor points, faulty distributor condenser, and various other items in the car electrical system can cause noisy operation. If the customer complains of noise in the receiver after it has been in use for some time, check the items mentioned as well as other parts of the car electrical system for poor connections, grounds, and other faults which may be responsible for the noise.

Circuit

The circuit consists of an antenna stage, a 78 R.F. stage, a 77 1st detector-oscillator stage, a 78 I.F. stage, a 75 dual diode-triode tube, which functions as a diode 2nd-detector and triode 1st audio stage, and a single 41 output stage. An 84 full wave rectifier is used in the power unit. The intermediate frequency is 262 K.C. The diode current establishes a drop across a resistor which is used as additional bias voltage for the R.F. and I.F. tubes giving automatic volume control action. Noise suppression between stations is obtained by the resistor in the cathode circuit of the 75 tube, the drop across which must be overcome before rectification in this tube begins. The manual volume control varies the audio voltage applied to the grid of the 75 tube.

A vibrator interrupts the current through the primary of the power transformer in the power unit. This, together with the turns ratio in this trans-

WELLS-GARDNER COMPANY

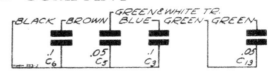

Fig. 10 Condenser Block Internal Wiring

former, results in the high voltage AC being present in the secondary of the transformer. The full wave rectifier tube, filter choke, and filter condensers convert this high voltage AC into high voltage DC for the plate and screen circuits.

Current for the receiver is obtained from the car storage battery. In Fig. 10 is shown the condenser block internal wiring.

Voltages at Sockets

In the following chart are given the voltages at the sockets with all the tubes in, all units connected, and the set in operating condition, but with no signal being received. The antenna should be grounded.

A thousand ohm-per-volt meter of 0-250 volt range is required for the plate and screen voltages.

Lower ranges will be necessary for the grid and heater voltages. It is not absolutely necessary to have a high resistance meter for the heater or "A" battery reading.

These voltages will vary with variations in receivers, tubes, test equipment used, and "B" eliminator output voltage.

Type of Tube	Function	Across Heater	Plate to Cathode	Screen to Cathode	Grid to Cathode	Normal Plate M.A
78	R. F.	6.1	182	80	3.0	7.0
77	1st Det. and Osc.	6.1	178	77	5.0	1.3
78	I. F.	6.1	182	80	3.0	7.0
75	2nd Det. 1st Audio	6.1	70(3)		1.4(2)	.35
41	Output	6.1	172.5	176.5	12.5(4)	16.0
84	Rect.	6.1	205			17.5 per plate

(1) Cathode to Ground
(2) Subject to Variation
(3) Triode Plate to Cathode
(4) Read Across 400 Ohm Resistor, R13

Misalignment or mistracking of condensers generally manifests itself in broad tuning and lack of volume at portions or all of the broadcast band. The receivers are all properly aligned at the factory with precision instruments and realignment should not be attempted unless all other possible causes of the faulty operation have first been investigated and unless the service technician has the proper equipment. A signal generator that will provide accurately calibrated signals over the broadcast band and accurately calibrated signals at and around 262 K.C. the intermediate frequency and an output indicating meter are desirable.

First set the signal generator at approximately 262 K.C. Connect the antenna lead from the generator to the control grid of the I.F. 78 tube, through a .05 mfd. condenser. The ground lead of the generator goes to the ground of the receiver. Turn the rotor plates of the tuning condenser completely out and keep the signal weak enough to prevent A.V.C. action. Note from Fig. 9 that the second I.F. transformer is self tuned and cannot be adjusted. Adjust the frequency of the signal generator until the output meter shows maximum output. The intermediate frequency setting of the generator is then correct, although it may be a very small percentage higher or lower than 262 K.C.

Next connect the signal lead from the signal generator to the grid of the 1st detector tube through a

.05 mfd. condenser. Then adjust the two intermediate frequency condensers for maximum output. One of the I.F. condenser screws is reached through the hole on the top of the 1st I.F. assembly can. The other I.F. condenser screw is reached from the bottom of the sub-panel through a hole at the bottom of this assembly.

Now set the signal generator for a signal of exactly 1400 K.C. The antenna lead from the generator is, in this instance, connected to the antenna lead of the receiver. Connect the flexible drive shaft to the chassis if it has been disconnected. As explained previously, the dial scale should be at the low frequency end and stop when the rotor is completely in mesh. Then turn the station selector knob until the dial scale is at 1400 K.C.

Then adjust the three trimmer condensers on the gang tuning condenser for maximum output, adjusting the oscillator section first.

Next, set the signal generator for a signal of 600 K.C. and adjust the oscillator 600 K.C. trimmer. The adjusting screw for this condenser is reached through a hole in the back wall of the sub-panel.

A non-metallic screwdriver is necessary for this adjustment. Turn the tuning condenser rotor until maximum output is obtained. Then turn the rotor slowly back and forth over this setting, at the same time adjusting the 600 K.C. trimmer screw until the highest output is obtained.

Then set the signal generator again for a signal of 1400 K.C. and check the adjustment of the tuning condenser trimmers at this frequency for maximum output.

ZENITH RADIO CORP.

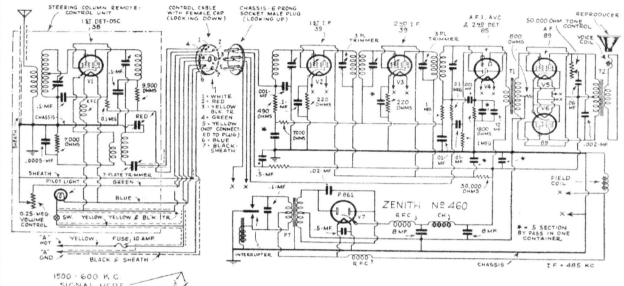

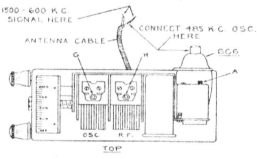

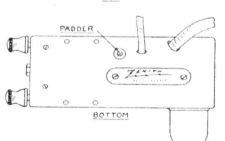

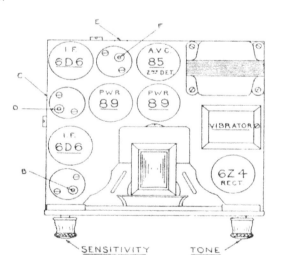

TUBE OPERATING VOLTAGES

Position		Tube	Ef	Ek	Eg^1	Eg^2	Eg^3	Ep
1st Detector		6C6	4.8	6.5	0	6.5	120	150
1st I.F. Amp.		6D6	5.3	10.5	*	10.5	103.5	165
2nd I.F. Amp.		6D6	5.3	10.5	*	10.5	103.5	165
2nd Detector		85	5.3	8.	0	—	—	156
P. P. Audio	{	89	5.3	17.	0	17.	165	165
		89	5.3	17.	0	17.	165	165

f—Filament. k—Cathode. g^1—Control grid. g^2—Suppressor grid. g^3—Screen grid. p—Plate.
*Depends on applied signal strength. All voltages measured from indicated points to ground.

ZENITH RADIO CORP.

Test Procedure

In the event that trouble develops it is advisable to first inspect the battery and antenna. A battery with a defective cell or in a run down condition will supply insufficient voltage to the receiver with a serious drop in efficiency. Check it for voltage and specific gravity.

Next inspect the antenna. The metal windshield moulding may have cut the insulation and shorted the wire. A continuity test will quickly determine its condition. In the case of under car systems inspect the insulators closely, since corrosion or road dirt is likely to create high leakage to the car frame.

If the receiver is entirely inoperative the fuse should be examined. It is contained in an insulated holder at the "Hot" battery terminal. Be sure to replace the spaghetti insulator over the fuse if necessary to change it. The next important step is to very carefully check the tubes both in the control head and speaker chassis. This has been found to be the most common cause of service in an auto receiver. The extreme vibration to which the tubes are subjected will occasionally develop a short in the elements in spite of the precautions that have been taken in their construction. A loud hum and lack of sensitivity can usually be attributed to a defective 6C6. Microphonic howl can be traced to the 89's. Replacement is recommended for such complaints, since the average tube checker will not show up this condition. An intermittent cutting out accompanied with rasping and other noises will usually be found in either of the 6D6's. The chassis may be taken out for inspection by simply removing the cable plug and three round-head hexagon nuts on the front of the case.

If further inspection indicates that the difficulty lies in the parts or wiring, a voltage reading at all sockets should be taken. They should coincide closely with the values given in the table. It is also advisable to check for continuity in the I.F. transformers both at the control head and in the speaker chassis.

Where the set lacks volume or sensitivity check the power output tubes and the overall alignment of the I.F., R.F. and padder adjustments as specified under "Alignment." Always make certain that the volume and sensitivity controls are in maximum position when making a service inspection.

Alignment

Every Zenith Automobile receiver is balanced on an accurate, temperature controlled crystal oscillator before leaving the factory and unless a part is changed or the calibration has shifted, the adjustments should not be tampered with. Where it is absolutely necessary, however, a good test oscillator capable of delivering a modulated signal at 1500, 600 and 485 K.C. will be required.

To balance the I.F. circuit remove the cap and lead from the grid of the 6C6 tube in the control head and attach the test oscillator to the grid and to ground. Set to 485 K.C. and first adjust the primary I.F. trimmer shown (A) in Figure 1. Next trim the secondary (B). Now turn the plate trimmer (C) on the side of the chassis base to resonance, with a No. 4 Spintite wrench. Its (2nd I.F.) transformer is directly above the adjustment. Set the screw (D) in the top of the transformer shield to resonance also. The third I.F. transformer is adjusted through a hole in the rear of the chassis and also on top of the transformer indicated at E and F. This completes the I.F. circuit. Replace the grid lead on the 6C6 and screw the metal cap back in position.

Next attach the test oscillator to the antenna and ground lead of the control head and set it to 1500 K.C. Remove the control head cover and set the variable condenser trimmers (G and H) to a point where the 1500 K.C. signal is loudest at that frequency on the dial. Then set the test oscillator to 600 K.C. and rock the dial slowly at that frequency; at the same time turn the padder condenser adjusting screw. This trimmer is reached by removing the button plug on the bottom of the control head. The adjustments should be gone over twice to insure greater precision. Provided the test equipment is dependable, the entire receiver will now be in proper alignment and the calibration very accurate.

Undercutting Insulation

The commutator bars of all generator armatures are insulated from each other by mica or a bakelite composition known as **micarta**. This insulation between the bars should be undercut about 1/32 inch in depth. (See Figs. 23 and 24.) When renewing brushes in a generator with an undercut armature, it is necessary to sand the brushes to a good seat to prevent noisy operation and arcing.

If an armature in service is found with the commutator worn, grooved or with a rough and burned surface, showing high insulation leakage between the commutator bars, it should be placed in a lathe and the commutator turned down. This work should be done carefully, as the surface of the commutator must be concentric with the armature shaft to insure proper performance. Before placing the armature in the lathe, remove any burrs or foreign material that may have collected in the center hole of the armature shaft. Turn the armature at a reasonably high speed and use a fine feed and a very sharp tool.

When the commutator is turned down, undercut the mica between the copper bars to 1/32 inch, keeping the slot rectangular in shape and the edges free from the insulating material.

There are several undercutting machines on the market which can be used for this purpose. In the absence of a machine, the work may be accomplished with a hack saw blade, after

having ground off the sides of its teeth until it will cut a slot slightly wider than the insulating material. (See Fig. 24.) The final assembly of a typical generator in its frame is shown in Fig. 25.

After the undercutting operation remove burrs and smooth off the commutator with No. 00 sandpaper. With the use of air, blow out all loose particles between the commutator bars after sanding.

OAK "B" ELIMINATOR

A PHOTOGRAPH and a schematic diagram of the new "B" eliminator produced by the Oak Manufacturing Company are shown. The type 180 eliminator has an output of 180 volts at 40 ma. when excited from a 6-volt D.C. source; under these conditions, it draws 2.3 amperes. The eliminator is provided with intermediate taps. The eliminator is equipped with a dual vibrator, feeding a push-pull primary circuit, giving a true alternating current input to an 84 rectifier. The efficiency of this unit is approximately 60% and secures its excitation from the leakage flux.

OAK MFG. CO.

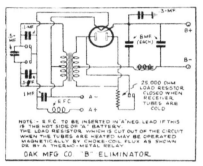

OAK MFG CO. "B" ELIMINATOR

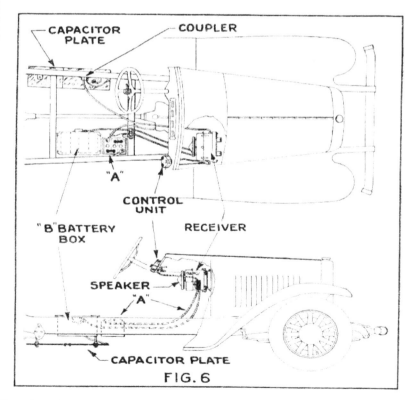

FIG.23

FIG.24

FIG.25

FIG. 6

You Servicemen

Make Me PROVE That I Can Show You How to Make

MORE MONEY

Men! Now is the time to do some tall thinking and perhaps a bit of stiff work. Review your trade and service journals for the past few months and get a load of the new-fangled developments in Radio — A. V. C.; Silent A. V. C. Muters; Multifunction Tubes; Class B Push-Push Amplifiers; Universal A.C., D.C., and Battery Receivers; Automatic Tone Control; Mercury Vapor Rectifiers and the like.

Do you honestly know how all these complicated circuits and tubes work? Do you really know what makes the wheels go around? Can you honestly say that you are able to keep up with these modern improvements?

If not, let me tell you straight from the shoulder that unless you buckle up and get down to brass tacks, learn the fundamentals of Radio, a new crop of Radio servicemen are about to bundle you out of your servicing job.

Stop Guessing. Know What You Are Doing

There are three major factors which confront every modern serviceman. You must consider the following facts if you want to go on.

1.

You must have a sound fundamental knowledge of Radio theory and Radio receiver design practice. If you are not grounded in the basic dope you can never keep abreast of new developments.

2.

You must have simple but effective testing and servicing equipment. You can't select them on anyone's say so. You must know about everything available and do your own choosing. This requires knowledge. Servicing is two-thirds knowledge and one-third testing equipment. It is just as necessary to know how your test equipment works as to know how a Radio set works.

3.

You must have a carefully selected service technique

which you should follow. It must be a technique that will stand the test of time; that will not be made obsolete by new developments. You must study the technique of servicing so you can go from effect to cause — by this I mean that a certain squeal or growl, a touch of the grid of a vacuum tube means something more to you than a noise.

This Plan Will Make More Money For You

Here is a fact that you know as well as I do. When you get stuck on a service job, all tangled up in the diagnosis of the trouble, what do you do? Naturally you start out from the beginning — make a fresh start. Why not apply the same principle to your background of Radio knowledge? Let me prove to you that I can give you a fresh start for a lasting association in servicing and one that will make more money for you.

MAIL THE COUPON
Get My Sample Lesson and School Catalog

My home-study training has helped hundreds of servicemen qualify for better jobs. Get my school catalog — see the facts for yourself — what my training covers, what it has done for others. Read my Money-Back Agreement. My catalog and sample lesson are FREE. See how thorough, how practical I've made my training. Send the coupon. There's no obligation. Act now.

J. E. SMITH, Pres., National Radio Institute
Dept. 3HD8   Washington, D. C.

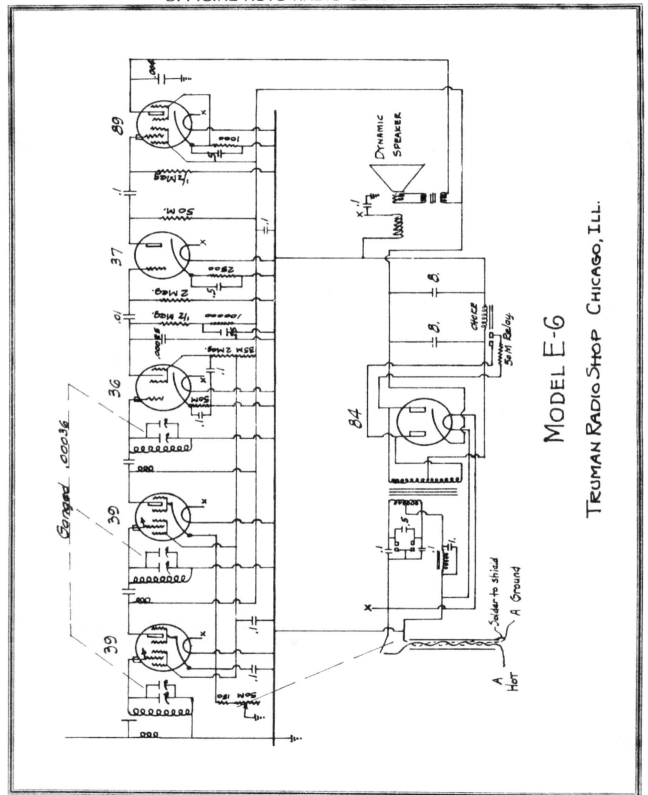

MODEL E-6

TRUMAN RADIO SHOP CHICAGO, ILL.

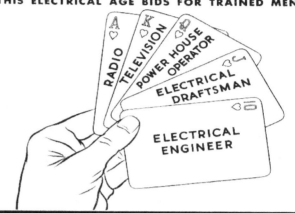

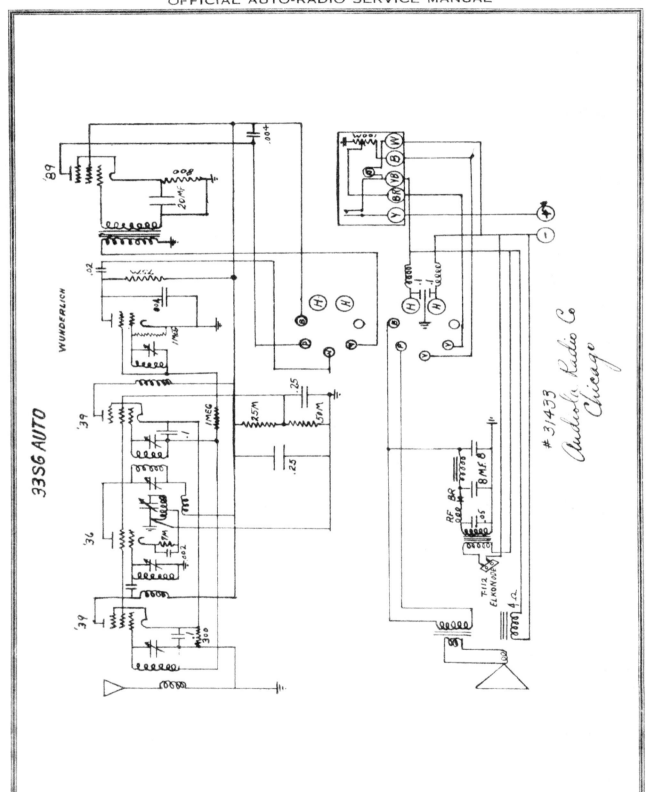

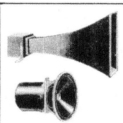

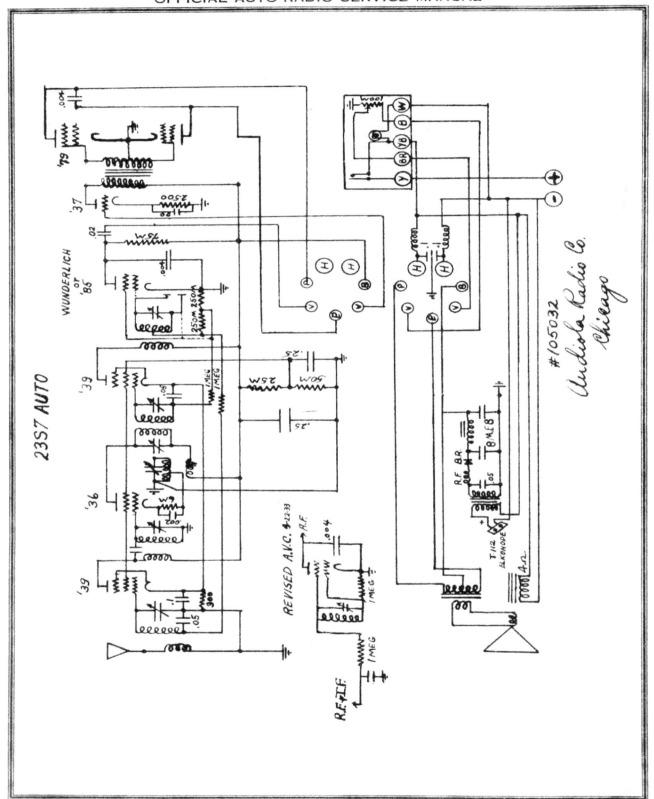

ALLIED RADIO CORP.

1930 BATTERY AUTO SET

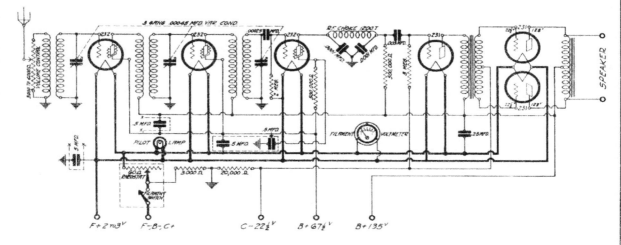

CIRCUIT—BATTERY RECEIVER—'30

ALLIED RADIO CORP.

MODEL A-S-30 S.G.

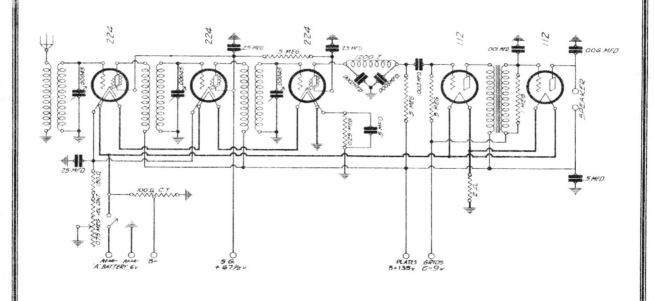

ATWATER KENT MFG. CO.

MODEL 81, 81-B, 81-C MOTOR CAR RADIO

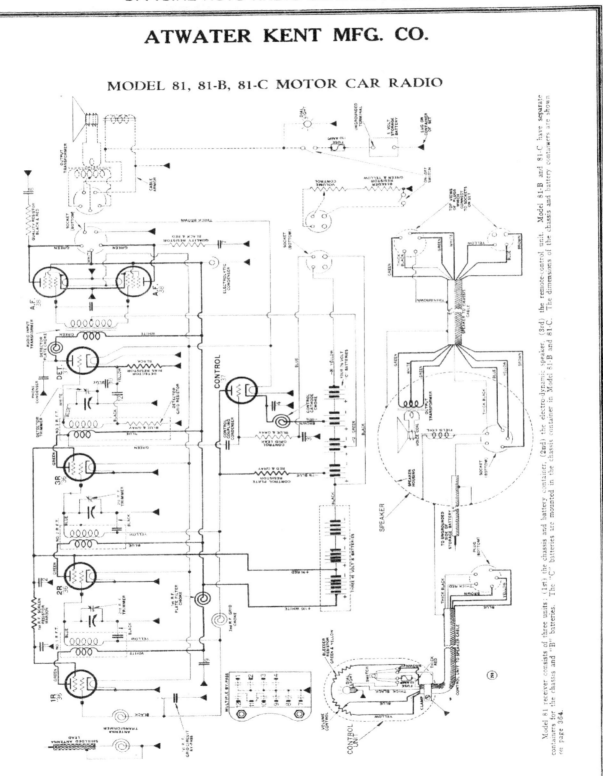

Model 81 receiver consists of three units: (1st) the chassis and battery container, (2nd) the electro-dynamic speaker (3rd) the remote-control unit. Model 81-B and 81-C have separate containers for the chassis and "B" batteries. The "C" batteries are mounted in the chassis container in Model 81-B and 81-C. The dimensions of the chassis and battery containers are shown on page 364.

ATWATER KENT MFG. CO.

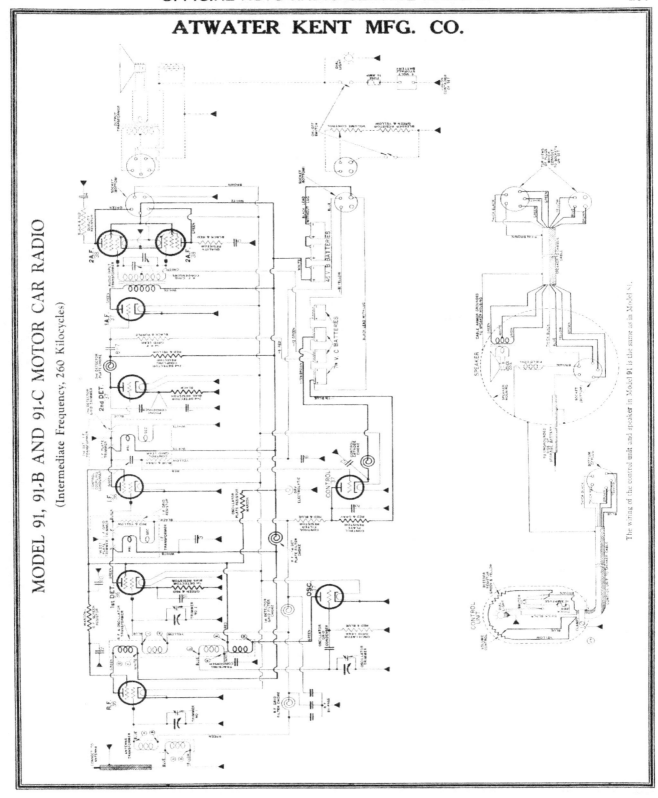

MODEL 91, 91-B AND 91-C MOTOR CAR RADIO

(Intermediate Frequency, 260 Kilocycles)

The wiring of the control unit and speaker in Model 91 is the same as in Model 81.

ATWATER KENT MFG. CO.

MODEL 424, 534

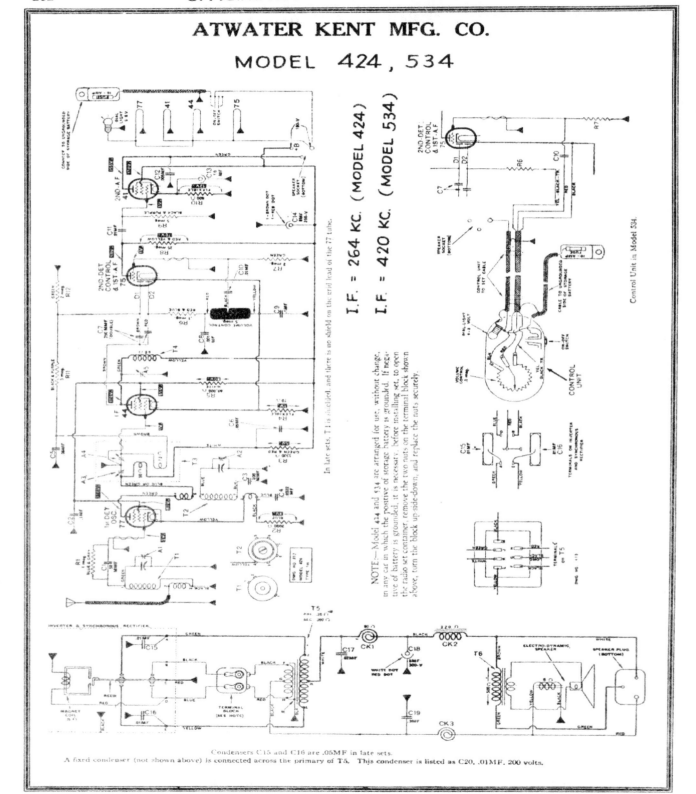

I.F. = 264 KC. (MODEL 424)

I.F. = 420 KC. (MODEL 534)

In late sets, T1 is shielded, and there is no shield on the grid lead of the 77 tube.

NOTE—Model 424 and 534 are arranged for use, without change, in any car in which the positive of storage battery is grounded. If negative of battery is grounded, it is necessary, before installing set, to open the radio set container, remove the two nuts on the terminal block shown above, turn the block up side-down, and replace the nuts securely.

Control Unit in Model 534.

Condensers C15 and C16 are .05MF in late sets.

A fixed condenser (not shown above) is connected across the primary of T5. This condenser is listed as C20, .01MF, 200 volts.

AUDIOLA RADIO CO.

MODEL 23-S-7

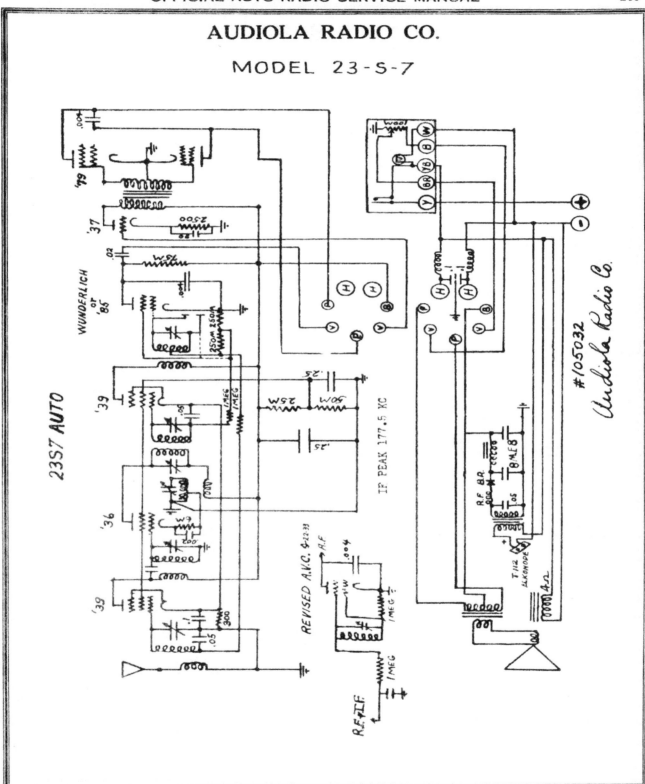

AUDIOLA RADIO CO.
MODEL 33A6, 33-S-7

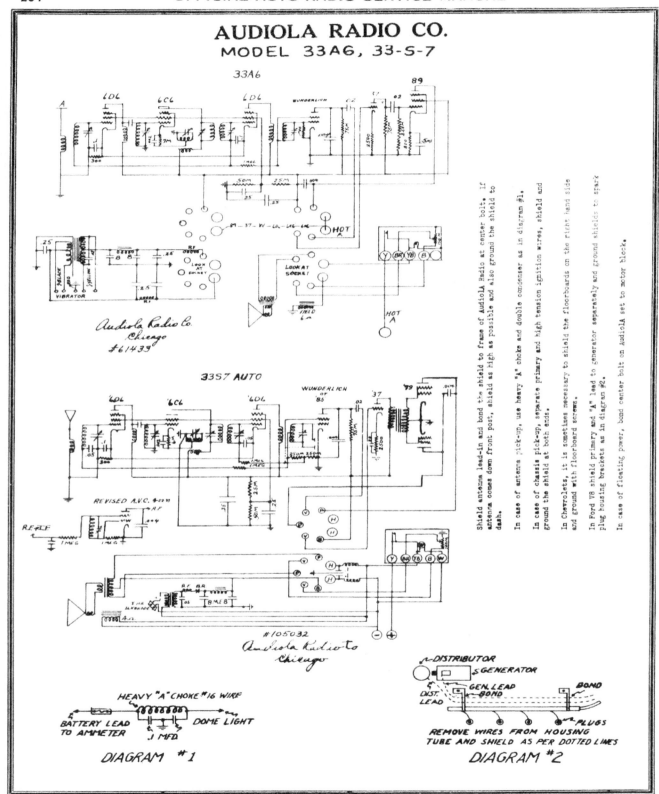

DIAGRAM #1

DIAGRAM #2

CENTURY RADIO PRODUCTS CO.

MODEL 7-38

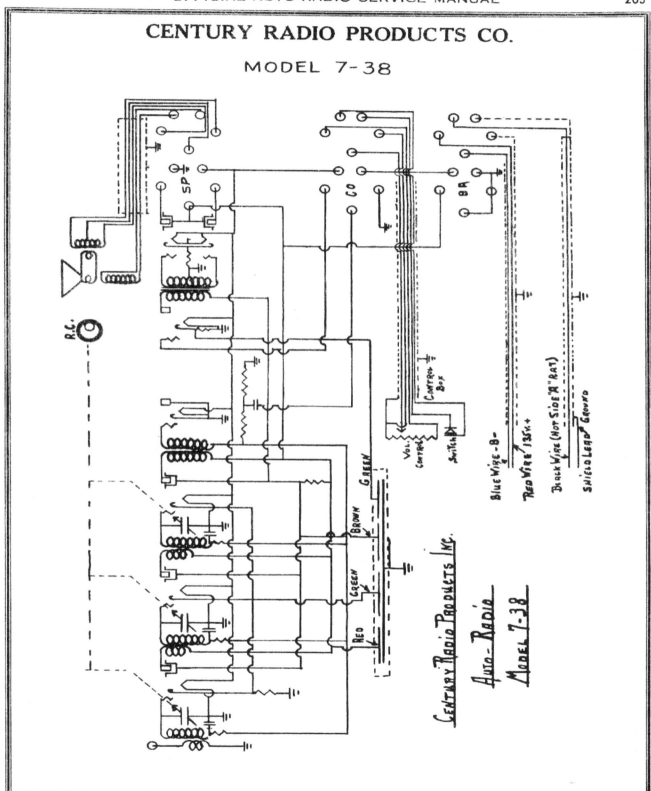

CONSOLIDATED INDUSTRIES PRODS. LTD.
ROAMIO MODEL 51

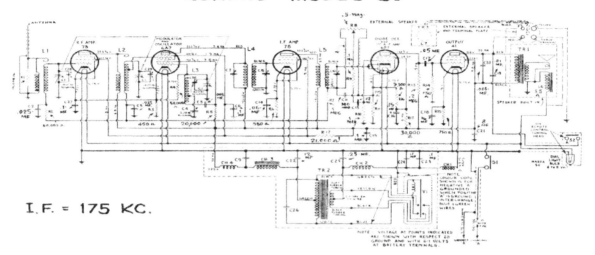

I.F. = 175 KC.

CROSLEY RADIO CORP.
MODEL 90 ROAMIO

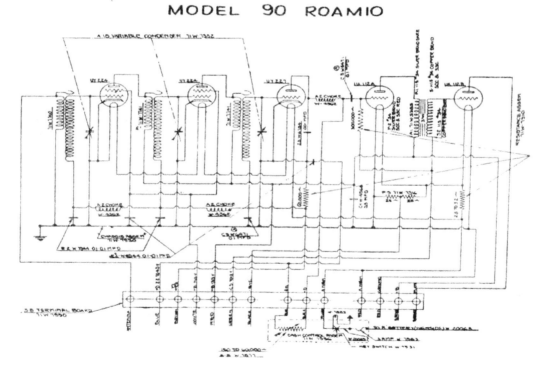

Filament Voltages		Control Grid Voltages	
R. F. and Detector Tubes	2.0	R. F. Tubes	2.5
A. F. Tubes	4.7	Detector Tube	3.0
		A. F. Tubes	12.0
Plate Voltages			
All Tubes but Detector	135	**Screen Grid Voltages**	
Detector Tube	22½	R. F. Tubes	90

CROSLEY RADIO CORP.
MODEL 91

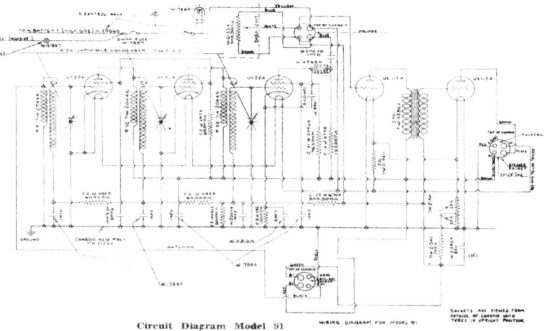

Circuit Diagram Model 91

WIRING DIAGRAM FOR MODEL 91

MODEL 92

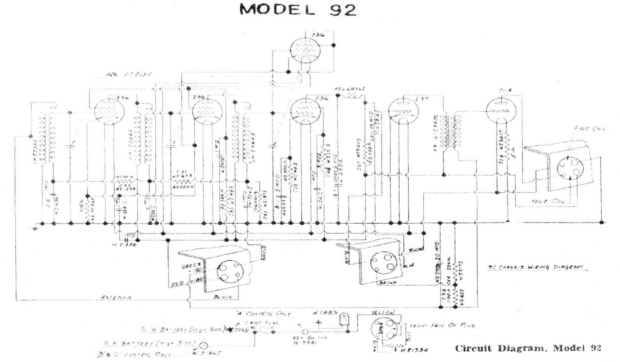

Circuit Diagram, Model 92

CROSLEY RADIO CORP.

Model 102

Specifications

Model 102 is a single unit five tube superheterodyne designed for operation from a six volt automobile battery. "B" voltage is furnished by the Crosley Syncronode which is built into the receiver. Intermediate frequency is 181.5 Kc.

Tubes and Voltage Limits

The following are the tubes and voltages measured with the receiver in operating condition but with no signal to the antenna circuit, and a battery voltage of 6.3 volts. All voltages are measured with a 250 volt D.C. voltmeter (1000 ohms per volt) from tube to chassis.

Tube	Position	Plate	Screen Grid	Cathode	Suppressor Grid	Filament
78	R. F. Amplifier	206	108	2.0	2.0	6.3
78	Oscillator Modulator	206	108	28.0	0	6.3
78	I. F. Amplifier	206	108	3.0	3.0	6.3
6B7	Detector and A. F. Amplifier	37	26	0		6.3
41	Output	198	206	16.0		6.3

Voltage limits should be plus or minus 15% of values given.

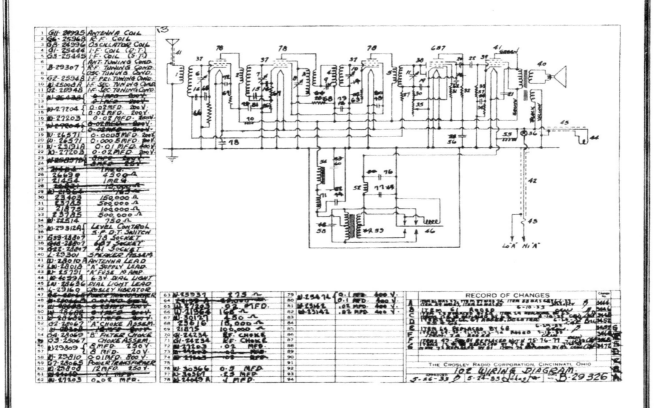

DELCO APPLIANCE CORP.
MODEL 3026

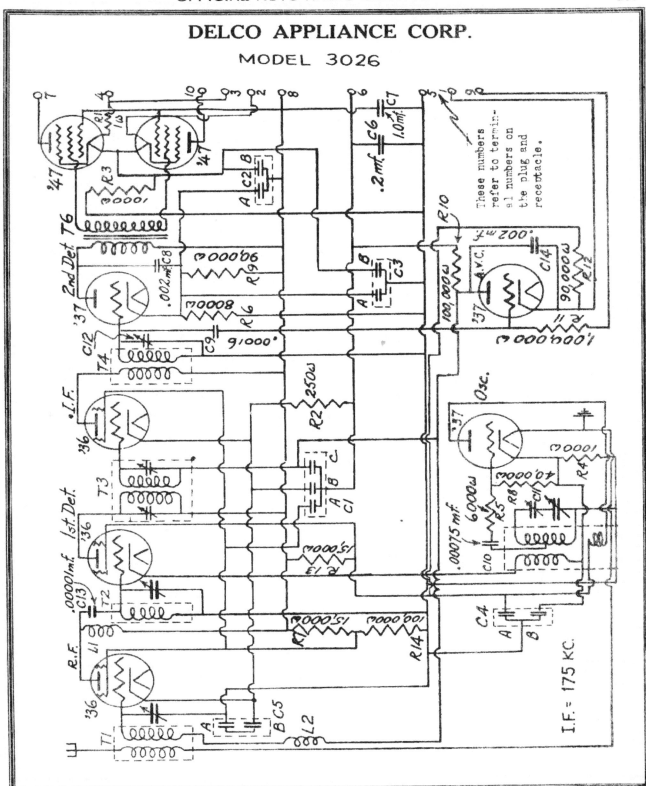

DETROLA RADIO CORP.

DETROLA "ROAD CHIEF"
WIRING DIAGRAM 5 TUBE T. R. F.

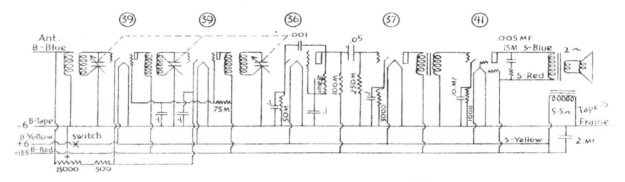

EMERSON RADIO & PHONOGRAPH CORP.
MODEL 678

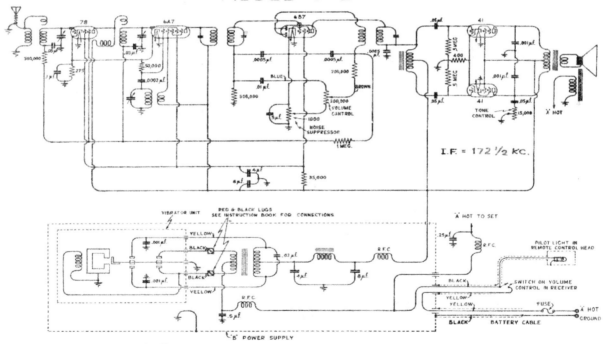

I.F. = 172 ½ KC.

Tube	Cathode to Ground	Screen Grid to Ground	Plate to Ground	Heater to Ground
78	3- 3.5V.	75-85V.	200-210V.	6V.
6A7	3- 3.5	75-85	200-210	6
6B7	3.5-4.5	75-85	200-210	6
41	14-18	200-210	190-200	6
41	14-18	200-210	190-200	6

Voltage across speaker field—6 volts.

GALVIN MFG. CORP.
MODEL 66

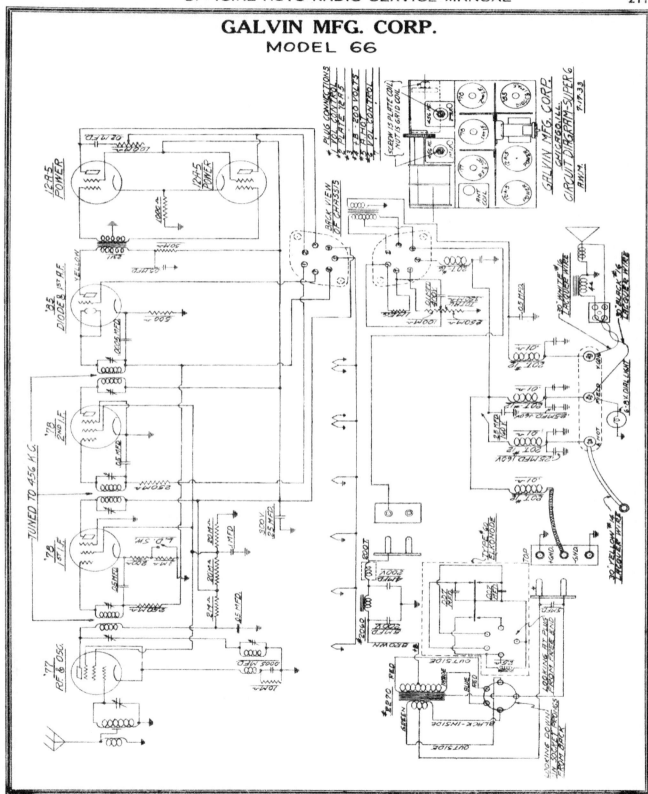

GALVIN MFG. CORP.
MODEL 77-A (SERIES A)

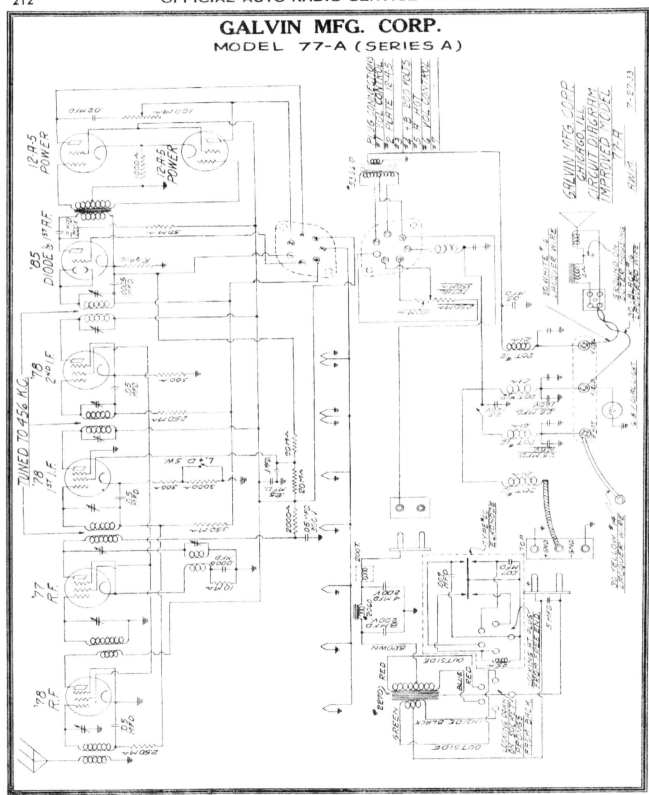

GALVIN MFG. CORP.
MODEL 77-A (SERIES B)

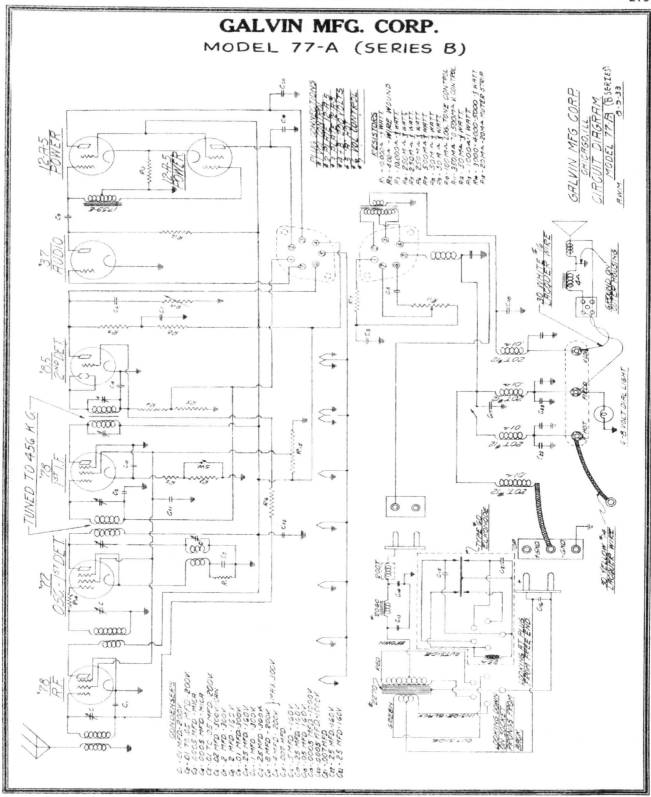

A. H. GREBE & CO.
MODEL 61

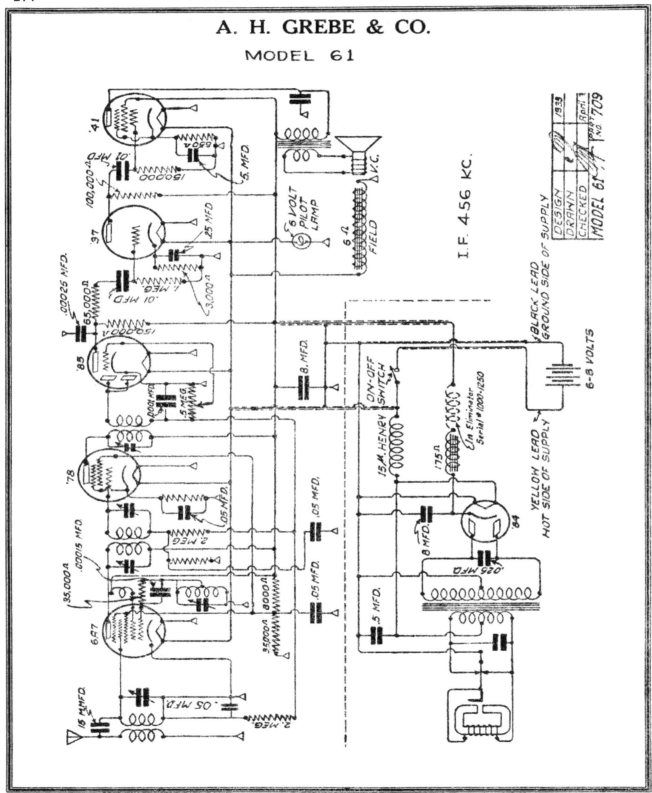

CHAS. HOODWIN CO.

AERO PENTODE AUTO RADIO
MODEL A — 1932   MODEL B — 1932

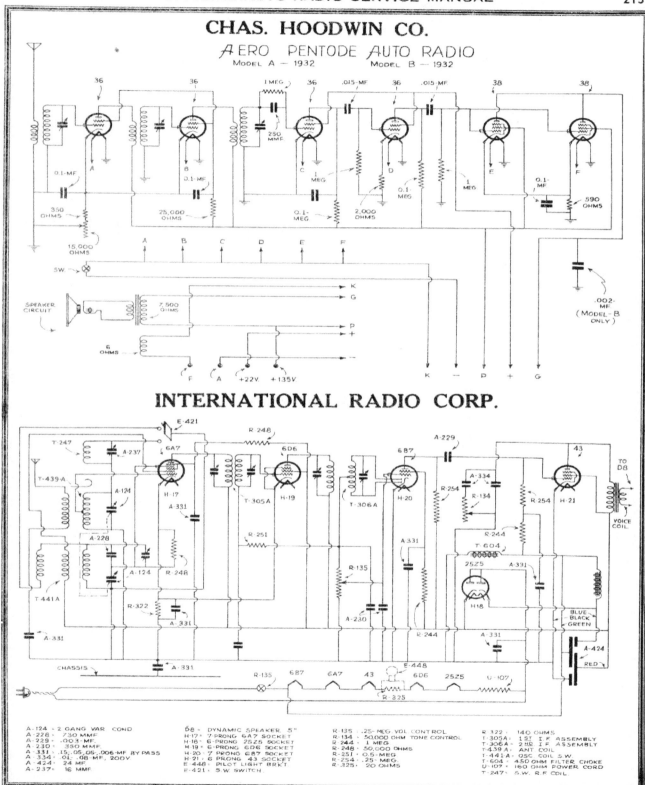

INTERNATIONAL RADIO CORP.

A-124 = 2 GANG VAR. COND	D8 = DYNAMIC SPEAKER, 5"	R-135 = .25-MEG. VOL CONTROL	R-322 = 140 OHMS
A-228 = 730 MMF.	H-17 = 7-PRONG 6A7 SOCKET	R-134 = 50,000 OHM TONE CONTROL	T-305A = 1ST I.F. ASSEMBLY
A-229 = .003-MF.	H-18 = 6-PRONG 2SZ5 SOCKET	R-244 = 1 MEG	T-306A = 2ND I.F. ASSEMBLY
A-230 = .350 MMF.	H-19 = 6-PRONG 6D6 SOCKET	R-248 = 50,000 OHMS	T-439-A = ANT. COIL
A-331 = .15;.05,.05,.006-MF BY PASS	H-20 = 7 PRONG 6B7 SOCKET	R-251 = 0.5-MEG	T-441-A = OSC. COIL S.W.
A-334 = .01-.08-MF, 200V	H-21 = 6 PRONG 43 SOCKET	R-254 = .25-MEG.	T-604 = 450 OHM FILTER CHOKE
A-424 = 24 MF	E-448 = PILOT LIGHT BRKT.	R-325 = 20 OHMS	U-107 = 160 OHM POWER CORD
A-237 = 16 MMF.	E-421 = S.W. SWITCH.		T-247 = S.W. R.F. COIL

P. R. MALLORY & CO.

STANDARD & TYPE C AUTO "B" ELIM.

ELKONODE RATING TABLE

Elkonode Type	Volts Output	For Receivers Requiring the Following Current in Milliamperes in the B Minus Lead at 200 V. on Signal		Elkonode Rated Output Watts	Storage Battery Drain In Amps.
		Without Voltage Dividers in Elim.	With 2 M.A. (100,000 Ohm) Voltage Divider in Elim.		
10	200	40—45	38—43	8.4	2.1
11	200	35—40	33—38	7.4	1.9
12	200	30—35	28—33	6.4	1.6
13	200	25—30	23—28	5.4	1.4
14	200	20—25	18—23	4.4	1.2

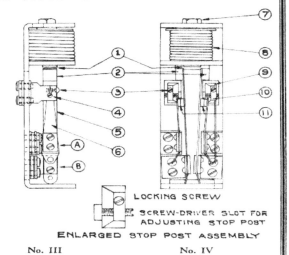

LOCKING SCREW

SCREW-DRIVER SLOT FOR ADJUSTING STOP POST

ENLARGED STOP POST ASSEMBLY

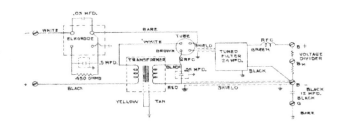

No. III
1. Air-gap
2. Reed counter weights
3. Stop-post Locking screw
4. Stop-post
5. Reed Spring Assm.

No. IV
6. Contact Spring Assm.
7. Coil mounting nut
8. Coil
9. Stop-post mounting block
10. Position contact spring behind stop-post head
11. Contact points

MELBURN RADIO MFG. CO.

MODEL 30

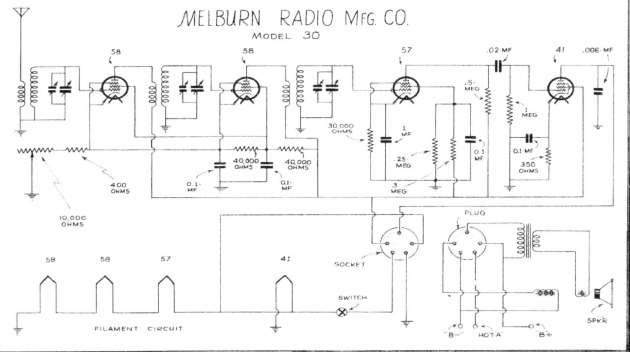

MELBURN RADIO MFG. CO.
MODEL 30

FILAMENT CIRCUIT

NOBILTT-SPARKS INDUSTRIES, INC.

ARVIN CAR RADIO MODEL 10A
RADIO FREQUENCY, SPEAKER & REMOTE CONTROL CIRCUIT DIAGRAM

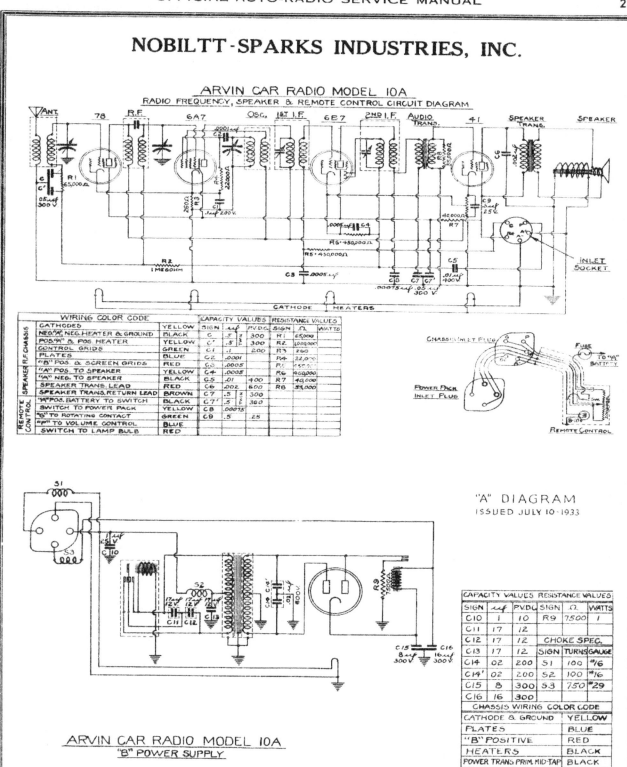

"A" DIAGRAM
ISSUED JULY 10-1933

ARVIN CAR RADIO MODEL 10A
"B" POWER SUPPLY

NOBILTT-SPARKS INDUSTRIES, INC.

SCHEMATIC CIRCUIT DIAGRAM
ARVIN CAR RADIO MODEL 20A
TYPE 1

DIAGRAM	ISSUE NO.	DATE
B	1	4-26-33
B	2	7-3-33

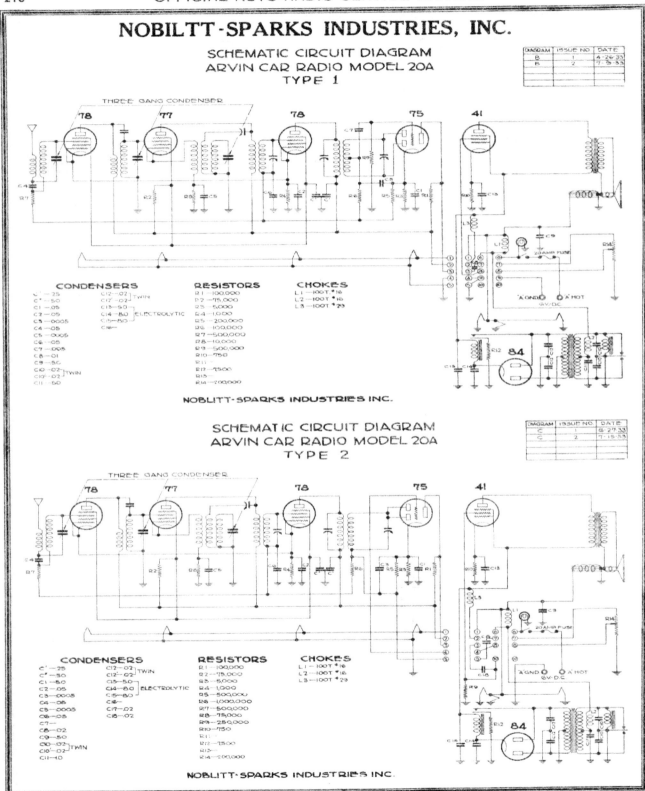

NOBLITT-SPARKS INDUSTRIES INC.

SCHEMATIC CIRCUIT DIAGRAM
ARVIN CAR RADIO MODEL 20A
TYPE 2

DIAGRAM	ISSUE NO.	DATE
C	1	6-27-33
C	2	7-15-33

NOBLITT-SPARKS INDUSTRIES INC.

NOBILTT-SPARKS INDUSTRIES, INC.

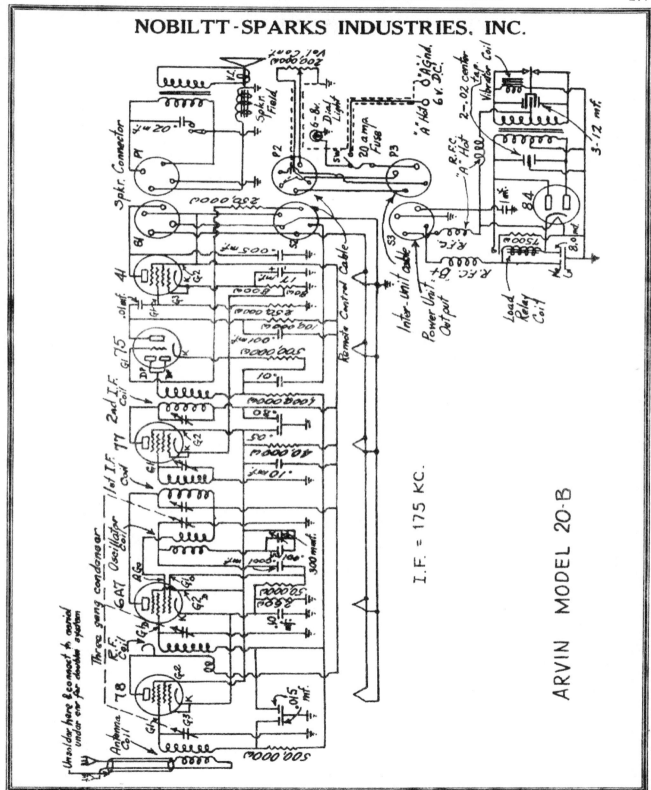

I.F. = 175 KC.

ARVIN MODEL 20-B

NOBILTT-SPARKS INDUSTRIES, INC.

SCHEMATIC CIRCUIT DIAGRAM
ARVIN CAR RADIO MODEL 30A
TYPE 1

DIAGRAM	ISSUE NO	DATE
B	1	4-26-33
B	2	7-8-33

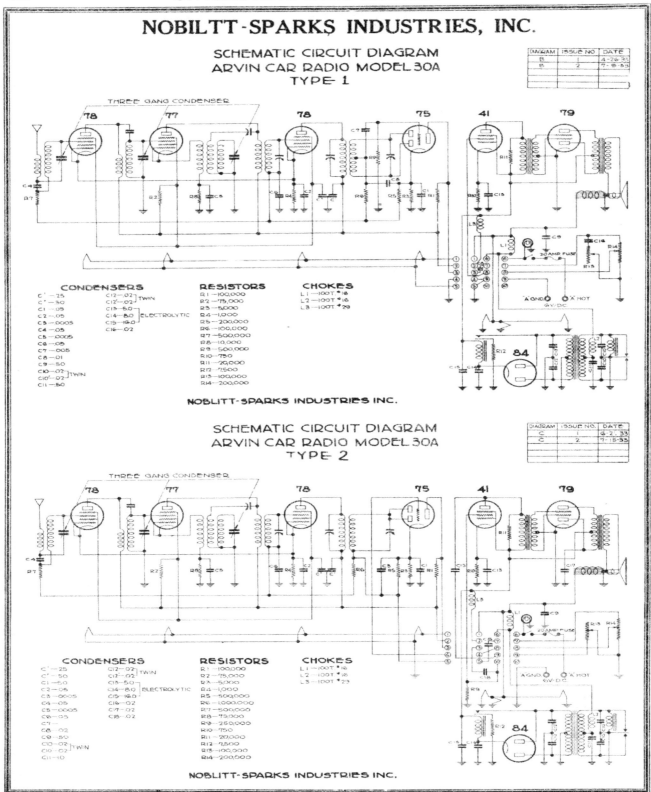

CONDENSERS

C¹ — 25	C12 — 02 TWIN
C¹ — 50	C12¹ — 02
C1 — 05	C13 — 5.0
C2 — 05	C14 — 8.0 ELECTROLYTIC
C3 — 0005	C15 — 16.0
C4 — 05	C16 — 02
C5 — 0005	
C6 — 05	
C7 — 005	
C8 — 01	
C9 — 50	
C10 — 02 TWIN	
C10¹ — 02	
C11 — 50	

RESISTORS

R1 — 100,000
R2 — 75,000
R3 — 5,000
R4 — 1,000
R5 — 200,000
R6 — 100,000
R7 — 500,000
R8 — 10,000
R9 — 500,000
R10 — 750
R11 — 20,000
R12 — 7500
R13 — 100,000
R14 — 200,000

CHOKES

L1 — 100 T #18
L2 — 100 T #18
L3 — 100 T #29

NOBLITT-SPARKS INDUSTRIES INC.

SCHEMATIC CIRCUIT DIAGRAM
ARVIN CAR RADIO MODEL 30A
TYPE 2

DIAGRAM	ISSUE NO	DATE
C	1	6-2-33
C	2	7-15-33

CONDENSERS

C¹ — 25	C12 — 02 TWIN
C¹ — 50	C12¹ — 02
C1 — 50	C13 — 50
C2 — 05	C14 — 8.0 ELECTROLYTIC
C3 — 0005	C15 — 16.0
C4 — 05	C16 — 02
C5 — 0005	C17 — 02
C6 — 05	C18 — 02
C7 —	
C8 — 02	
C9 — 50	
C10 — 02 TWIN	
C10¹ — 02	
C11 — 10	

RESISTORS

R1 — 100,000
R2 — 75,000
R3 — 50,000
R4 — 1,000
R5 — 500,000
R6 — 1,000,000
R7 — 500,000
R8 — 75,000
R9 — 250,000
R10 — 750
R11 — 20,000
R12 — 7,500
R13 — 100,000
R14 — 200,000

CHOKES

L1 — 100 T #18
L2 — 100 T #18
L3 — 100 T #29

NOBLITT-SPARKS INDUSTRIES INC.

PHILCO RADIO & TELEVISION CORP.
MODEL 10

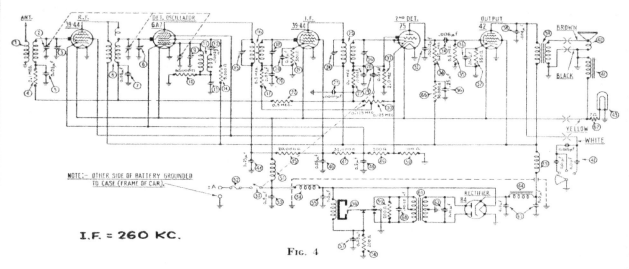

I.F. = 260 KC.

FIG. 4

MODEL 10 PARTS LIST

① Antenna Transformer	32-1220	㉑ Resistor (500 ohms)	6977	㊶ Field Coil Assembly	36-3130	㊴ Power Transformer	32-7098
② Tuning Condenser	31-1083	㉒ Resistor (500,000 ohms)	6097	㊷ Tone Control	30-4056	㊽ Condenser (.01 mfd.)	30-4051
③ 1st Padder (in tuning cond.)		㉓ Condenser (.00011 mfd.)	4519	㊸ Pilot Lamp	6608	㊾ Filter Condenser	30-2015
④ Resistor (100,000 ohms)	6099	㉔ Padder (prim. 2nd I.F.)	31-6008	㊹ Condenser (.25 mfd.)	04360	㊿ B Chokes	32-7038
⑤ Condenser (.05 mfd.)	30-4020	㉕ I.F. Transformer (2nd)	32-1237	㊺ Resistor (20,000 ohms)	6649	㊻ R. F. Chokes	32-1078
⑥ R.F. Transformer	32-1221	㉖ Padder (secondary 2nd I.F.)	31-6008	㊻ Condenser (.05 mfd.)	30-4020	㊼ Resistor (50,000 ohms)	4237
⑦ Condenser (.05 mfd.)	30-4020	㉗ Resistor (100,000 ohms)	6099	㊼ Resistor (32,000 ohms)	3525	㊿ Resistor (7 ohms)	5110
⑧ 2nd Padder (in tuning cond.)		㉘ Condenser (.00025 mfd.)	3082	㊽ Condenser (.5 mfd.)	30-4048	Spark Plug Resistors	33-1015
⑨ 3rd Padder (in tuning cond.)		㉙ Condenser (.01 mfd.)	30-4051	㊾ Resistor (200 ohms)	7217	Distributor Resistor	4546
⑩ Resistor (50,000 ohms)	6098	㉚ Vol. Control Assembly	38-5280	㊿ Resistor (100 ohms)	7838	Screw Type Resistor	4851
⑪ Oscillator Transformer	32-1222	㉛ Resistor (2,000,000 ohms)	33-1025	㊿ A Choke	32-1268	Interference Condenser	30-4007
⑫ Condenser (.00025 mfd.)	3082	㉜ Condenser (.00025 mfd.)	5858	㊿ 15 Amp. Fuse	7227	Dial	27-5022
⑬ Padder	040008	㉝ Resistor (250,000 ohms)	3768	㊿ Condenser (.5 mfd.)	30-4061	Studs	28-6036
⑭ Resistor (15,000 ohms)	6208	㉞ Condenser (.006 mfd.)	30-4024	㊿ Vibrator Choke	32-1259	Nuts (mounting)	W55
⑮ Padder (prim. 1st I.F.)	31-6007	㉟ Resistor (500,000 ohms)	6097	㊿ Condenser (.5 mfd.)	30-4061	Knobs	03534
⑯ I.F. Transformer (1st)	32-1236	㊱ Condenser (20 mfd.; 25 mfd.)	30-4065	㊿ Vibrator	38-5036	Battery Cable	38-5296
⑰ Resistor (500,000 ohms)	6097	㊲ Resistor (550 ohms)	6977	㊿ Condenser (.05 mfd.)	30-4039	Antenna Lead	38-5161
⑱ Padder (secondary 1st I.F.)	31-6007	㊳ Condenser (.006 mfd.)	30-4024	㊿ Resistor (200 ohms)	7217	Control Unit Assembly	42-5056
⑲ Condenser (.05 mfd.)	30-4020	㊴ Output Transformer	32-7102	㊿ Resistor (200 ohms)	7217	Acorn Nut	W521
⑳ Condenser (.5 mfd.)	30-4058	㊵ Cone and Coil	36-3020	㊿ Condenser (.00125 mfd.)	5886	Key	6091

I. F. TRANSFORMER AND PADDERS

A new style I. F. transformer complete with padders is used in the Model 10.

The padders are placed in the top of the shield can one above the other.

The primary padder is adjusted by means of the screw slot, accessible through the hole in the top of the shield can. The secondary padder is adjusted by means of the small hex nut, also accessible through the hole in the top of the shield. (See Figs. 1 and 2.)

The coil windings terminate in leads instead of terminals or lugs. The color scheme of the leads is given in Fig. 1.

If replacements are ever necessary, replace the entire coil assembly 38-5274 for the first I. F. stage and 38-5275 for the second I. F. stage. Neither the coil nor the padders will be furnished separately. Order only by the above numbers.

FIG. 1

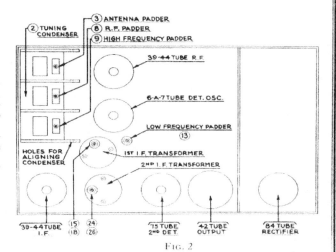

FIG. 2

PIERCE - AIRO, INC.
MODEL 54

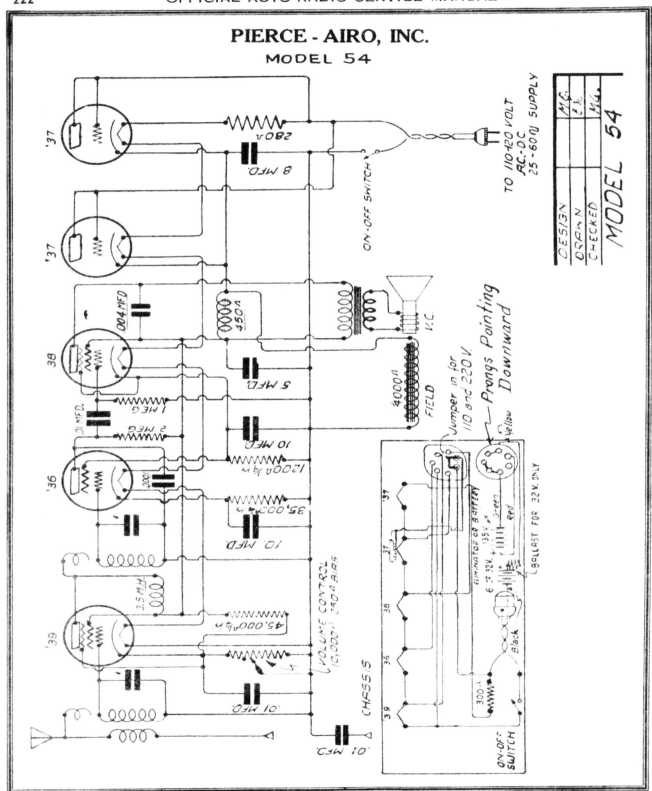

RADIO CHASSIS, INC.
MODEL V-6

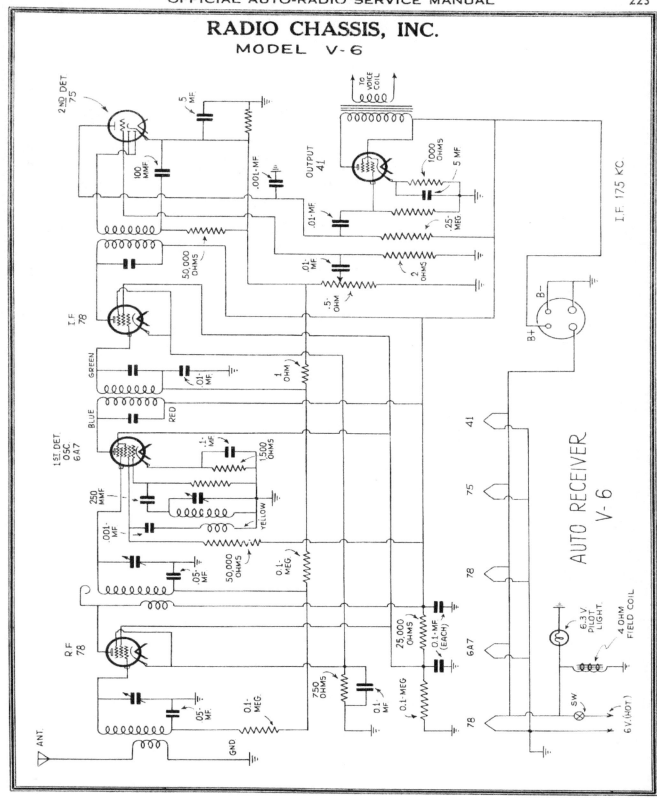

RCA-VICTOR, Inc.

MODEL M-30

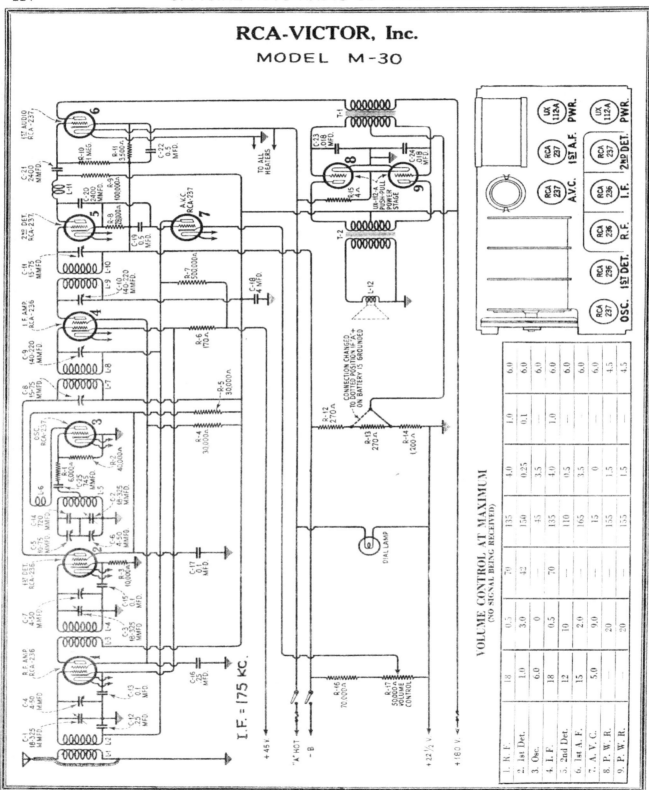

RCA-VICTOR, Inc.
MODEL M-32

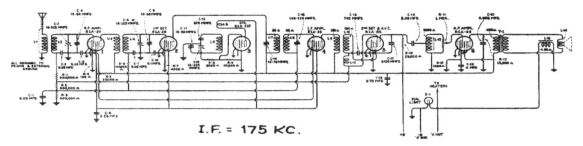

I.F. = 175 KC.

Line-up Capacitor Adjustments

The receiver must be removed from its metal case to permit correct adjustment of the line-up capacitors. After being removed, a grounded metal plate must be provided for the receiver to rest upon, otherwise the adjustments will be found to be incorrect when the assembly is returned to its metal case. After removal from its case and placing upon the metal plate, proceed as follows:

I. F. Line-up Capacitor Adjustment—The I. F. Amplifier uses two transformers, one being of the untuned variety and one having each of its windings tuned by means of two adjustable capacitors. Figure A shows the location of these capacitors.

(a) Procure a modulated oscillator giving a signal at 175 K. C. and having its output adjustable. A non-metallic screwdriver such as Stock No. 7065 is necessary together with an output meter.

(b) Remove the receiver from its case, place it in operation and connect the output of the oscillator between the control grid and ground of the first detector. Remove the oscillator tube and connect the output meter—preferably a thermo-galvanometer

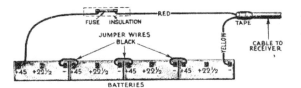

R. F. Line-up Capacitor Adjustment—The R. F., 1st detector and oscillator stages are aligned at 1400 K. C. A modulated oscillator giving a signal at 1400 K. C. a socket wrench and an output meter are necessary for correctly making these adjustments.

(a) Remove the receiver from its metal case and place on a grounded metal plate. Connect the tuning control and place in operation. Connect the output of the oscillator between antenna and ground. Connect the output meter across the voice coil of the loudspeaker.

(b) Place the oscillator in operation at 1400 K. C. and adjust its output so that a small deflection is obtained when the receiver volume control is at maximum and the dial set at 1400. Then adjust the three line-up capacitors until a maximum deflection is obtained. This is done by means of a socket wrench.

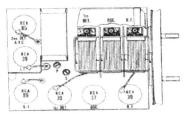

Location of Radiotrons and Line-up Capacitors

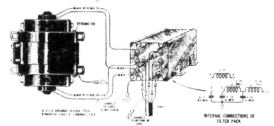

Plate Supply Unit Wiring

RADIOTRON SOCKET VOLTAGES

Radiotron No.	Cathode or Filament to Control Grid Volts	Cathode or Filament to Screen Grid Volts	Cathode or Filament to Plate Volts	Plate Current M. A.	Filament or Heater Volts
1. R.F. RCA-39	0.9	71	177	4.5	5.2
2. 1st Det. RCA-39	6.0	67	172	1.35	5.2
3. Osc. RCA-37			72	5.5	5.2
4. I.F. RCA-39	0.9	71	177	4.5	5.2
5. 2nd Det. and A.V.C. RCA-85			175	4.5	5.2
6. PWR. RCA-89	18	178	160	18.0	5.2

Voltages are those at which Radiotrons are operating and with no signal impressed on input

OTHER IMPORTANT VOLTAGES

Battery Voltage . . . . 6.0 Volts
Input to Dynamotor . . . . 5.75 Volts
Battery Drain . . . . 6.5 Amperes
Output from Dynamotor . . . . 178 Volts at 34.5 M.A.
Loudspeaker Field Drain . . . . 1.35 Amperes

SPARKS - WITHINGTON, INC.
MODEL AR-19

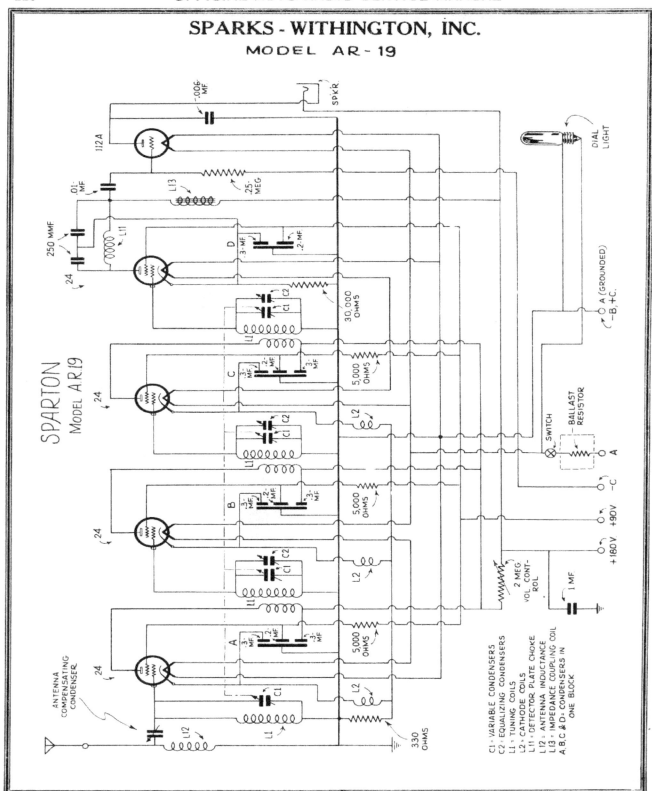

SPARKS - WITHINGTON, INC.
MODEL 40

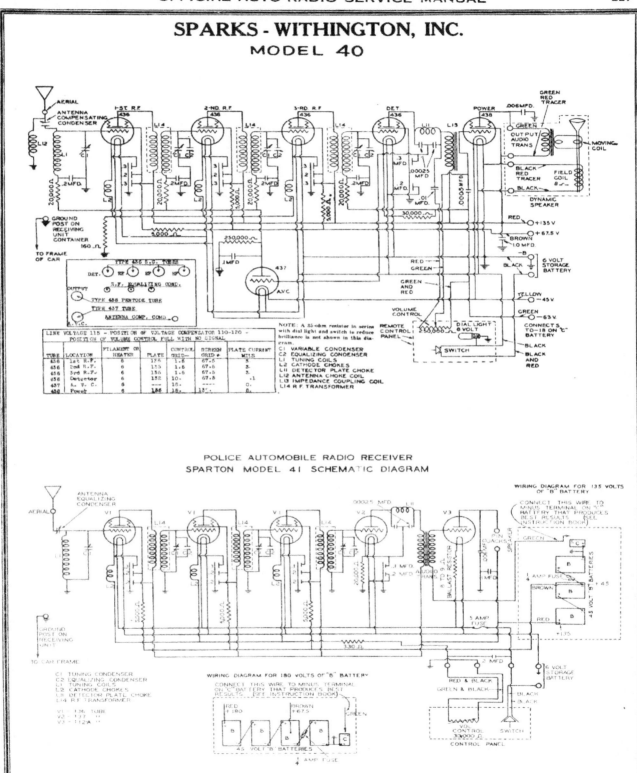

POLICE AUTOMOBILE RADIO RECEIVER
SPARTON MODEL 41 SCHEMATIC DIAGRAM

SPARKS - WITHINGTON, INC.
MODEL 43

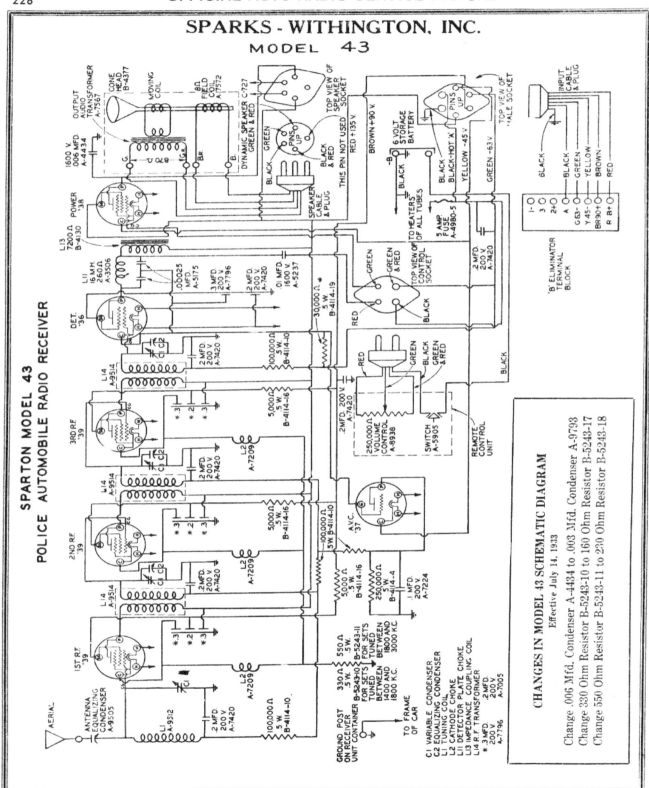

SPARTON MODEL 43
POLICE AUTOMOBILE RADIO RECEIVER

CHANGES IN MODEL 43 SCHEMATIC DIAGRAM

Effective July 14, 1933

Change .006 Mfd. Condenser A-4434 to .003 Mfd. Condenser A-9793
Change 330 Ohm Resistor B-5243-10 to 160 Ohm Resistor B-5243-17
Change 550 Ohm Resistor B-5243-11 to 230 Ohm Resistor B-5243-18

STEWART RADIO CORP.

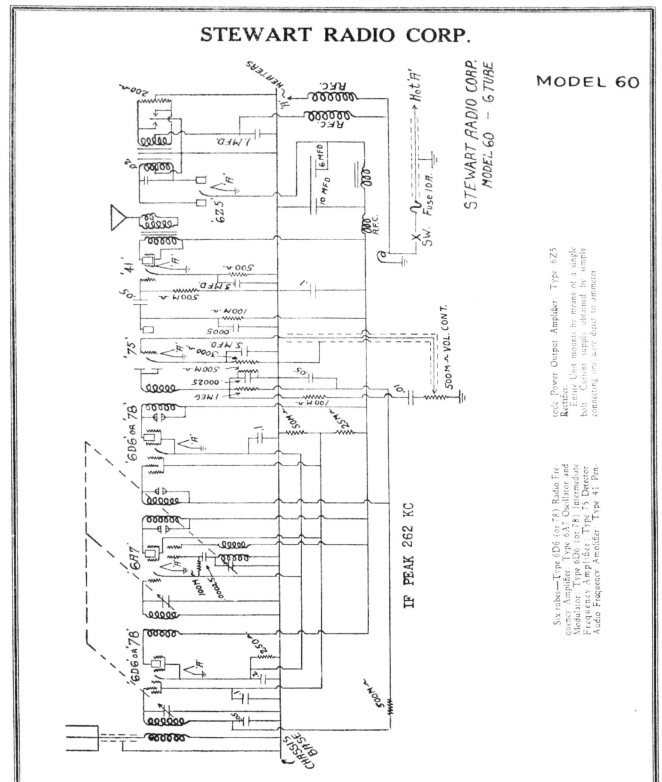

MODEL 60

STEWART-WARNER CORP.

Circuit Data for Stewart-Warner Chassis Series 108 and 108-X
Used on Models 10 to 20 Inclusive

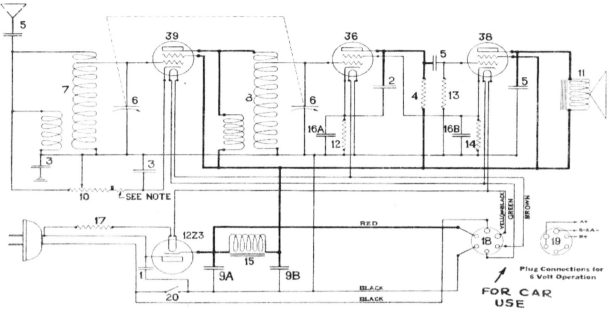

NOTE: In some receivers, a 140 ohm, ¼ watt carbon resistor, part 81646 is connected in series with the volume control; in other sets this resistor is built into the volume control.

LINE VOLTAGE 115 VOLTS A. C. ★ Voltage Table ★ VOLUME CONTROL FULL ON

Type of Tube	Tube Circuit	Filament to Condenser	Plate to Condenser	Screen Grid to Condenser	Cathode to Condenser
39	R. F.	** (See Note)	107	107	1.5
36	Det.	** (See Note)	1.3†	9	1.3
38	Output	** (See Note)	103	107	9
12Z3	Rect.	** (See Note)			122

IMPORTANT NOTE

★ These voltages will be obtained when the set is operated at 115 volts, 60 cycles A. C. For D. C. operation, voltages will be somewhat lower. All voltage readings have been taken between tube prongs and the variable condenser frame, *not the chassis*. The chassis cannot be used in this receiver as a reference point for voltage readings.
**Filament voltage readings will vary widely, depending upon the resistance of the A. C. voltmeter. With high resistance rectifier type meters, voltage readings will be approximately 6.3 for the detector and amplifier tubes, and 12.6 for the 12Z3 rectifier. With ordinary A. C. Voltmeters, readings will be very much less.
† This reading is obtained with a 30-volt scale, one thousand ohms per volt instrument. Higher resistance meters or higher scale readings will give greater voltage readings.

FRONT OF CHASSIS

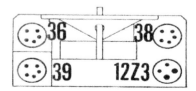

PARTS LIST

Diag. No.	Part No.	Description	List Price
1	67298	.01 mfd. 600 V cartridge condenser	$0.30
2	81158	.0001 mfd. mica condenser	.22
3	81630	.1 mfd. 400 V cartridge condenser	.30
4	81644	2.1 meg. ¼ W. carbon resistor	.20
	81646	140 ohm ¼ W. carbon resistor	.20
5	81657	.003 mfd. mica condenser	.35
6	81662	Variable condenser	3.00
7	81664	Antenna Coil	.90
8	81666	Detector Coil	1.20
9A 9B	81678	4 mfd. 150 V dry electrolytic condensers (in one unit)	1.35
10	81679	250,000 ohm volume control and switch	1.50
11	81680	Speaker	5.00
12	81681	29,000 ohm ¼ W. carbon resistor	.20
13	81682	1.1 meg. ¼ W. carbon resistor	.20
14	81683	1600 ohm ½ W. carbon resistor	.20
15	81694	Filter choke	2.00
16A 16B	81698	5 mfd. 20 V dry electrolytic condensers (in one unit)	1.20
17	81785	Power cord assembly	1.30
18	81834	Battery cable socket	.10
19	81861	6 volt battery cable	.90
	81863	12 volt battery cable	
	81865	32 volt battery cable	1.15
20	......	Switch on back of 81679	

PARTS NOT LISTED ON DIAGRAM

81885	Bronze tuning knob	.35
81886	Bronze volume control knob	.35
81887	Gold tuning knob	.35
81888	Gold volume control knob	.35
81889	Silver tuning knob	.35
81890	Silver volume control knob	.35
81891	Book model knob	.35
81824	Antenna reel	.25
81841	Antenna reel clip	.02
81712	Bronze receiver housing	4.00

STEWART-WARNER CORP.
MODEL R 112 & 112-1

STEWART-WARNER MODEL 1121 AUTO RADIO (R 112 CHASSIS)

I.F. 456 K.C.

VOLTAGE TABLE

Battery Voltage 60 — Volume Control Full On

Tube Type	Position in Circuit	Filament Voltage	Plate Voltage	Screen Grid Voltage	Cathode (Bias) Voltage
6A7	1st Det. and Osc.	5.5	144	70	1.4
78	I.F	5.5	144	70	2.0
75	2nd Det.	5.5	60	—	1.0
41	Output	5.5	142	144	9.0
84	Rect.	5.5	—	—	179

All D. C. voltages measured with respect to ground, using high resistance voltmeter of 1000 ohms per volt. Readings will vary, depending upon voltage range of meter, being higher for higher range instruments. This variation is most marked for second detector plate voltages.

TUBE LOCATIONS

VIBRATOR

DO NOT ATTEMPT TO ADJUST VIBRATOR OR GUARANTEE IS VOIDED. REPLACE WITH PART NO. 81371 IF NECESSARY.

41 - 6 Z 4 OR 84
OUTPUT   RECT.

75
2nd DET.

78
I.F.

6 A 7
1st DET. & OSC.

UPPER SHAFT-TUNING
LOWER SHAFT-VOLUME

FRONT OF SET

STROMBERG-CARLSON TELEPHONE MFG. CO.
MODEL 31 POLICE SET

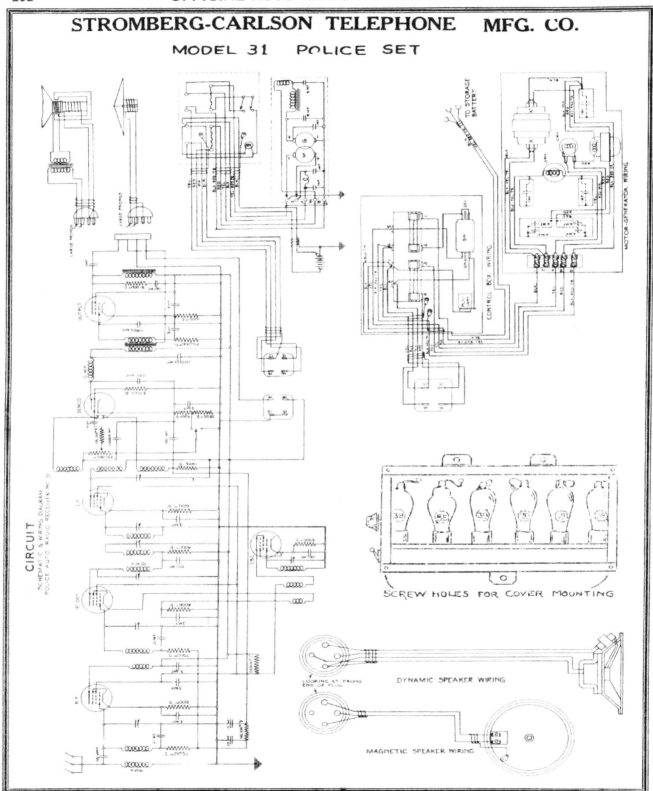

CIRCUIT
SCHEMATIC & WIRING DIAGRAM
POLICE AUTO RADIO RECEIVER NO. 31

TO STORAGE BATTERY

CONTROL BOX WIRING

MOTOR-GENERATOR WIRING

SCREW HOLES FOR COVER MOUNTING

LOOKING AT PRONG END OF PLUG

DYNAMIC SPEAKER WIRING

MAGNETIC SPEAKER WIRING

STROMBERG-CARLSON TELEPHONE MFG. CO.
MODEL 33

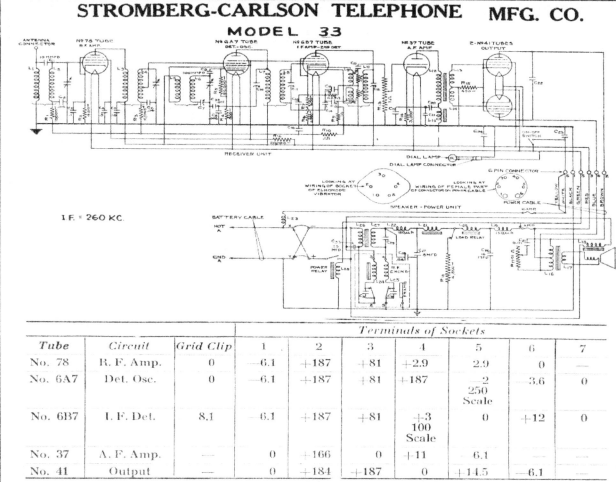

I. F. = 260 KC.

Tube	Circuit	Grid Clip	Terminals of Sockets						
			1	2	3	4	5	6	7
No. 78	R. F. Amp.	0	−6.1	+187	+81	+2.9	2.9	0	—
No. 6A7	Det. Osc.	0	−6.1	+187	+81	+187	−2 250 Scale	−3.6	0
No. 6B7	I. F. Det.	8.1	−6.1	+187	+81	+3 100 Scale	0	+12	0
No. 37	A. F. Amp.	—	0	+166	0	+11	−6.1	—	—
No. 41	Output	—	0	+184	+187	0	+14.5	−6.1	—

Note—These readings are made with the positive pole of the storage battery grounded. If the negative is grounded, the heater voltages will naturally be reversed. These voltages will vary slightly from the average given due to tolerances in resistors, variations in tubes, battery voltage differences, etc.

I. F. Aligners.

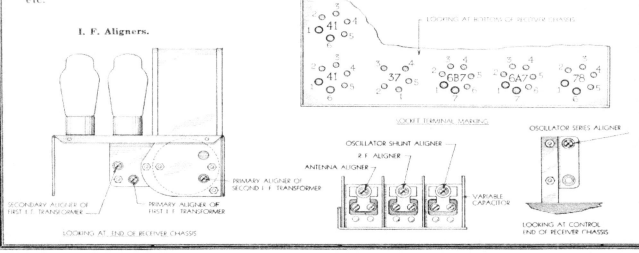

SECONDARY ALIGNER OF FIRST I.F. TRANSFORMER

PRIMARY ALIGNER OF FIRST I.F. TRANSFORMER

PRIMARY ALIGNER OF SECOND I.F. TRANSFORMER

LOOKING AT END OF RECEIVER CHASSIS

SOCKET TERMINAL MARKING

ANTENNA ALIGNER

OSCILLATOR SHUNT ALIGNER

R.F. ALIGNER

VARIABLE CAPACITOR

OSCILLATOR SERIES ALIGNER

LOOKING AT CONTROL END OF RECEIVER CHASSIS

TRANSFORMER CORPORATION OF AMERICA
MODEL 100

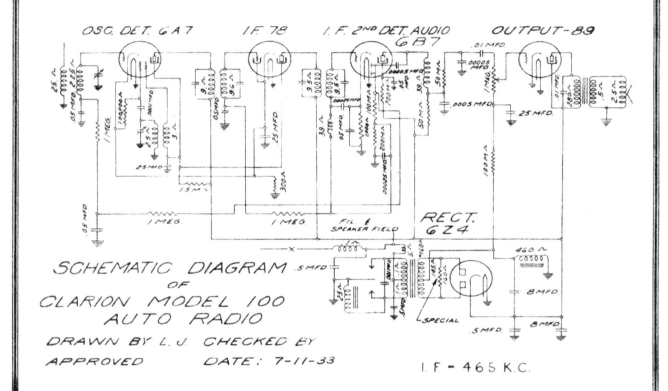

SCHEMATIC DIAGRAM
OF
CLARION MODEL 100
AUTO RADIO

DRAWN BY L. J. CHECKED BY

APPROVED   DATE: 7-11-33

I. F. = 465 K.C.

SOCKET VOLTAGE ANALYSIS OF MODEL

100-AR USING A 1000 OHM PER VOLT METER

No.	Stage	Tube	Ef	Ep	Eg	Ek	Esg	Esug	Ip	Ep—O	Eg—O	Ip—O
1	Osc.- Det . .	6A7	6	185	.1	3	83		4.6	81	.05	1.7
2	I. F.	78	6	185	.1	3	102	0	7.5			
3	I. F. 2nd Det. Audio .	6B7	6	58	.05	2.3	45		2.2	d .1		
4	Output. . . .	89	6	190	.05	0	194	0	18			
5	Rectifier . .	6Z4	6	p208		185			p18			

O — Oscillator.   p — Per Plate.
Volume Control — Full On.   d — Diode Plate.
Battery Voltage — 6 Volts.

UNITED AMERICAN BOSCH CORP.

Model 140 A

WIRING DIAGRAM

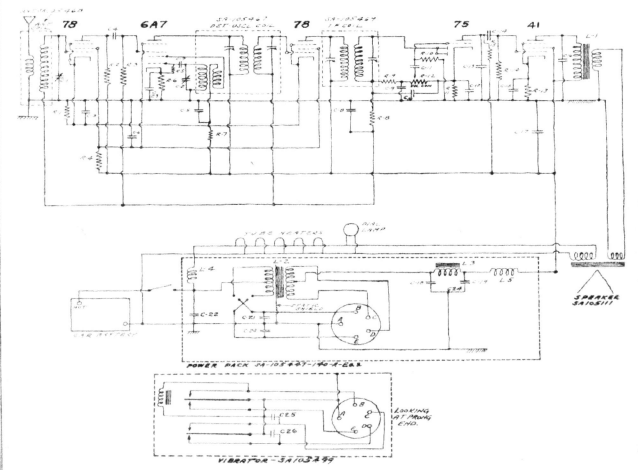

INTERMEDIATE FREQUENCY - 456 K.C.

To remove the tubes, it is necessary to remove the five thumb nuts which hold the front of the set in place, and lay the front cover, with speaker attached, on the floor-boards of the car. The three tubes enclosed by the shield alongside of the tuning condenser are then removed by placing a wire noose over the tube and pulling out. The other tubes can be pulled out directly with the fingers.

The vibrator may be withdrawn from the set when the front cover of the set is removed.

UNITED AMERICAN BOSCH CORP.

MODEL 150
TYPE 2

I.F. = 175 KC.

Total drain on the car battery: 6.1 amperes maximum

Output: 2.2 watts

Intermediate frequency: 175 kilocycles

"B" voltage: 190 volts or more under set load
                                 (with 6 volt storage battery)

UNITED AMERICAN BOSCH CORP.

MODEL 160

I.F. = 175 KC.

Total drain on the car battery: 6.1 amperes maximum

Output: 2.2 watts

Intermediate frequency: 175 kilocycles

"B" voltage: 190 volts or more under set load
                                 (with 6 volt storage battery)

UNITED MOTORS SERVICE
MODEL 4036; B-O-P

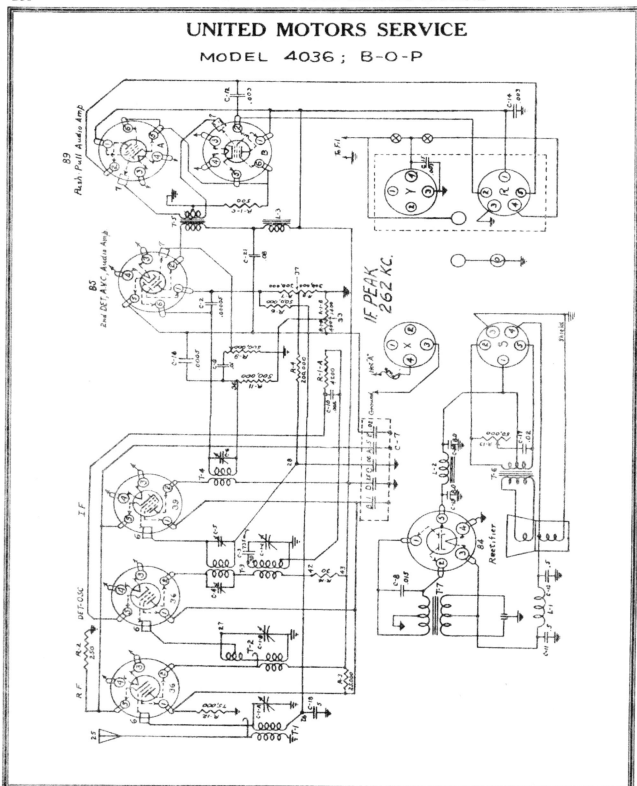

UNITED MOTORS SERVICE

MODEL 4037

UTAH RADIO PRODUCTS CO.
AUTO "B" ELIMINATOR

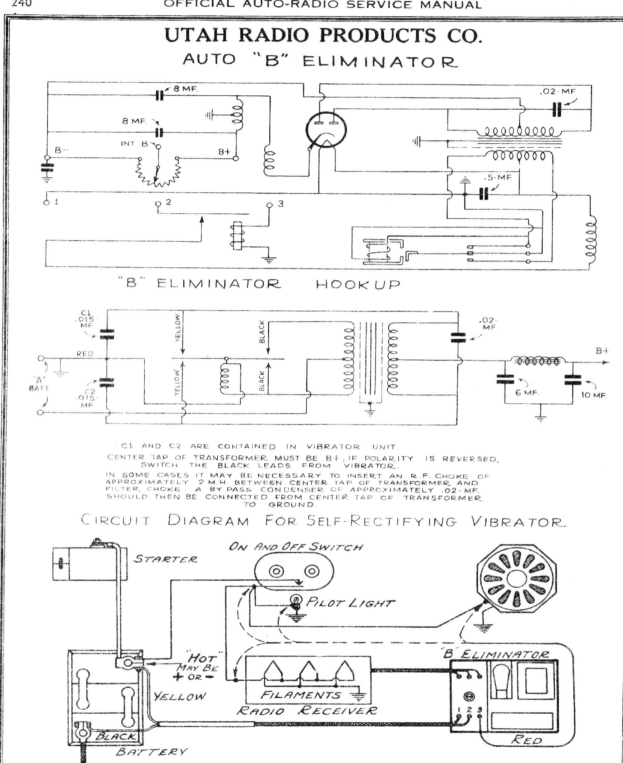

"B" ELIMINATOR HOOKUP

C1 AND C2 ARE CONTAINED IN VIBRATOR UNIT.
CENTER TAP OF TRANSFORMER MUST BE B+, IF POLARITY IS REVERSED, SWITCH THE BLACK LEADS FROM VIBRATOR.
IN SOME CASES IT MAY BE NECESSARY TO INSERT AN R.F. CHOKE OF APPROXIMATELY 2 M.H. BETWEEN CENTER TAP OF TRANSFORMER AND FILTER CHOKE. A BY PASS CONDENSER OF APPROXIMATELY .02 MF. SHOULD THEN BE CONNECTED FROM CENTER TAP OF TRANSFORMER TO GROUND.

CIRCUIT DIAGRAM FOR SELF-RECTIFYING VIBRATOR

WHOLESALE RADIO SERVICE CO., INC.

MODEL L-22

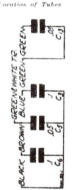

I.F. = 262 KC.

NOTE
ALL RESISTOR VALUES IN OHMS.
ALL CONDENSER VALUES IN MICROFARADS UNLESS OTHERWISE INDICATED.
ALL CONDENSERS WITH ASTERISKS (*) IN BLOCK.

Location of Tubes

Condenser Block—Internal Wiring

Voltages at Sockets

In the following chart are given the voltages at the sockets with all the tubes in, all units connected, and the set in operating condition, but with no signal being received. The antenna should be grounded.

A thousand ohm-per-volt meter of 0-250 volt range is required for the plate and screen voltages.

Lower ranges will be necessary for the grid and heater voltages. It is not absolutely necessary to have a high resistance meter for the heater or "A" battery reading.

These voltages will vary with variations in receivers, tubes, test equipment used, and "B" eliminator output voltage.

Type of Tube	Function	Across Heater	Plate to Cathode	Screen to Cathode	Grid to Cathode	Normal Plate M.A.
78	R.F.	6.1	192	80	3(3)	7.0
77	1st Det. and Osc.	6.1	178	77	5(3)	1.3(3)
78	I.F.	6.1	192	80	3(3)	7.0
75	2nd Det. 1st Audio	6.1	70(3)		1.4(3)	.35
41	Output	6.1	172.5	176.5	12.5(3)	16.0
84	Rect.	6.1	205			17.5 per plate

(1) Cathode to Ground
(2) Subject to Variation
(3) Triode Plate to Cathode
(4) Read Across 400-Ohm Resistor, R13

Circuit

The circuit consists of an antenna stage, a 78 R.F. stage, a 77 1st detector-oscillator stage, a 78 I.F. stage, a 75 dual diode-triode tube, which functions as a diode 2nd-detector and triode 1st audio stage, and a single 41 output stage. An 84 full wave rectifier is used in the power unit. The intermediate frequency is 262 K.C. The diode current establishes a drop across a resistor which is used as additional bias voltage for the R.F. and I.F. tubes giving automatic volume control action. Noise suppression between stations is obtained by the resistor in the cathode circuit of the 75 tube, the drop across which must be overcome before rectification in this tube begins. The manual volume control varies the audio voltage applied to the grid of the 75 tube.

A vibrator interrupts the current through the primary of the power transformer in the power unit. This, together with the turns ratio in this trans-

former, results in the high voltage AC being present in the secondary of the transformer. The full wave rectifier tube, filter choke, and filter condensers convert this high voltage AC into high voltage DC for the plate and screen circuits.

Current for the receiver is obtained from the car storage battery. In Fig. 11 is shown the condenser block internal wiring.

THE RUDOLPH WURLITZER MFG. CO.
MODEL A-60

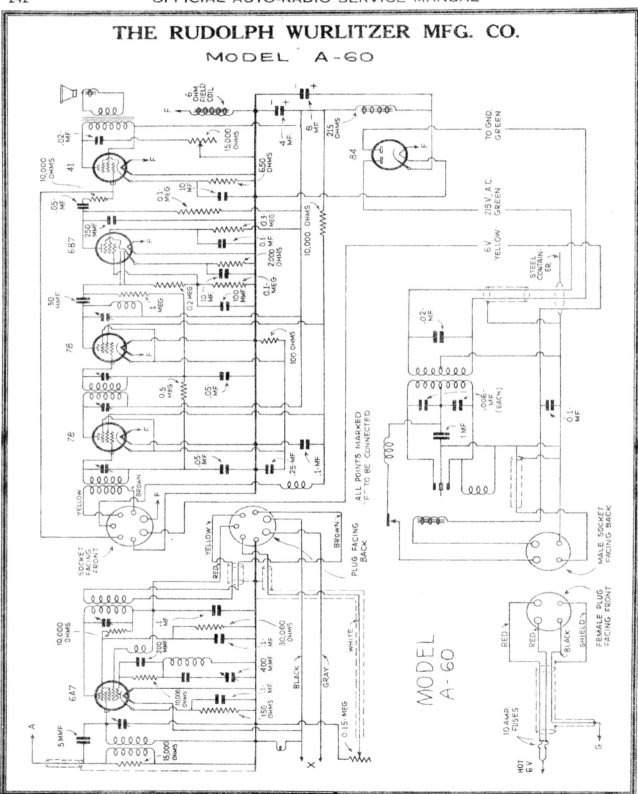

THE RUDOLPH WURLITZER MFG. CO.

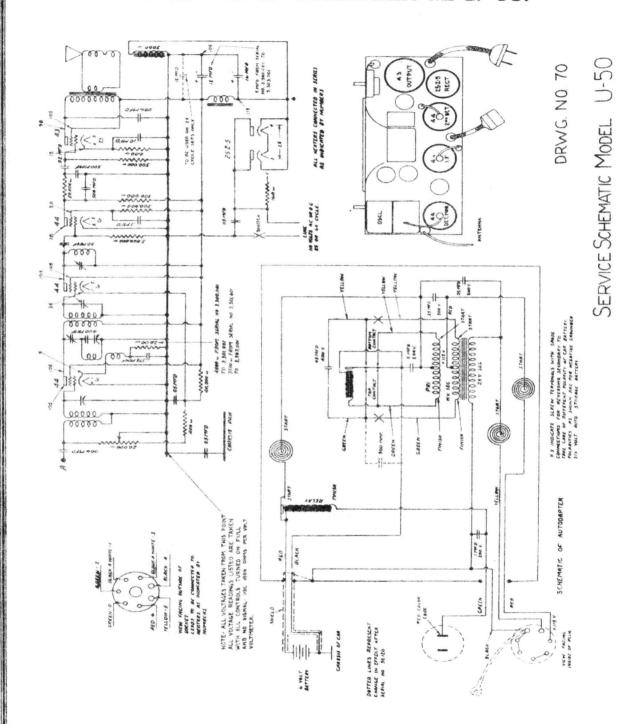

SERVICE SCHEMATIC MODEL U-50

DRWG. NO. 70

THE RUDOLPH WURLITZER MFG. CO.

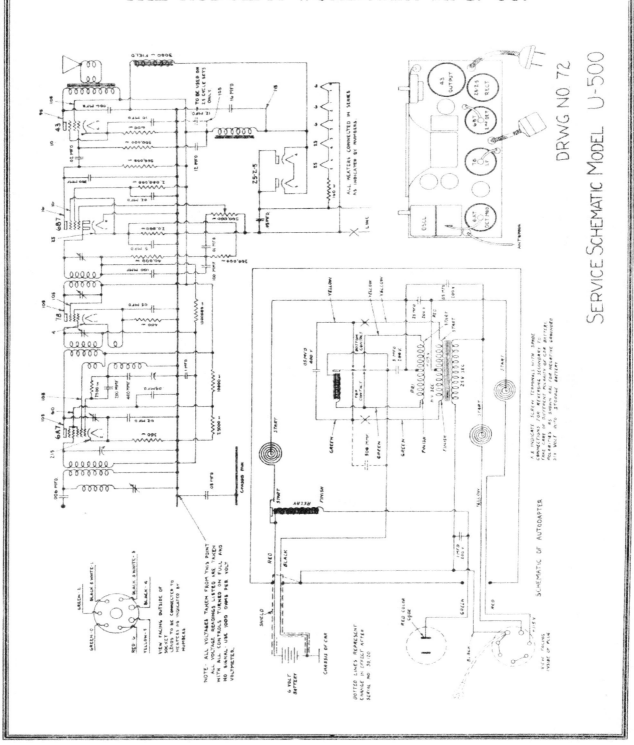

SERVICE SCHEMATIC MODEL U-500

DRWG NO 72

www.ingramcontent.com/pod-product-compliance
Lightning Source LLC
Chambersburg PA
CBHW080402060326
40689CB00019B/4108